THE SELF-HOSTED AI

BUILDER'S GUIDE

Deploy, Secure, and Operate

Private LLMs, AI Agents,

Voice Pipelines, and

Document Intelligence with Docker

Lin Song

Paperback ISBN: 978-1-970482-07-2

Published by VectorHarbor Books

Trademarks

Technical and security disclaimer

This book provides technical information for educational and informational purposes. Although reasonable care has been taken in its preparation, software, services, security practices, and provider requirements change over time. The information is provided without warranties, and no configuration is guaranteed to be suitable for a particular environment.

Commands can change systems, expose services, consume paid resources, or cause data loss. Review each command, protect credentials, maintain current backups, and test in a controlled environment before using it with important data or public traffic.

Third-party projects, models, images, providers, and security guidance change over time. Consult current primary documentation and applicable terms, licenses, and policies. You remain responsible for system administration, security, privacy, legal compliance, and the consequences of operating the software. To the fullest extent permitted by applicable law, the author and publisher disclaim liability for loss, damage, or disruption arising from the use of, or reliance on, the information in this book.

Contents

Introduction: Build an AI Platform You Control

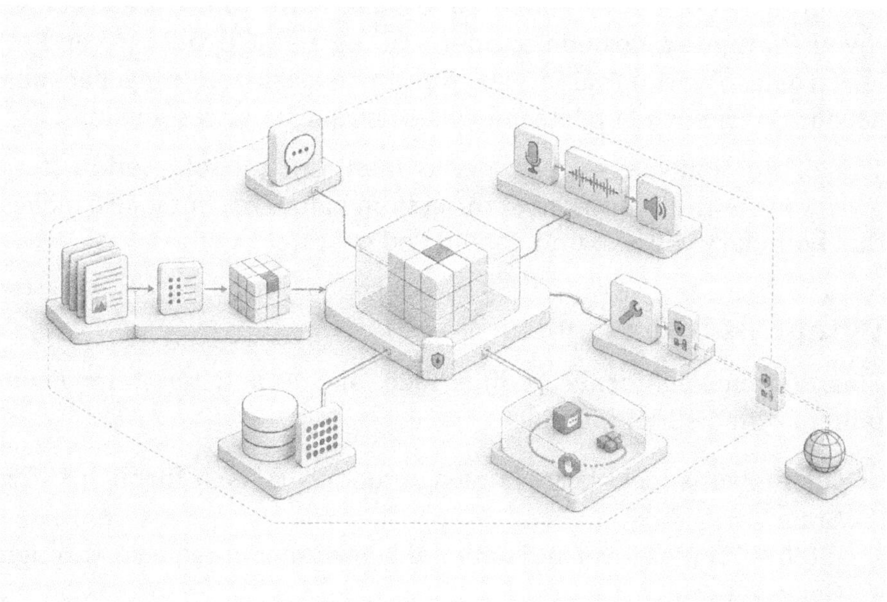

This book guides you through building a practical artificial intelligence platform on infrastructure you control. You can run local language models and a private browser chat interface, prepare documents for retrieval, transcribe and generate speech, and connect compatible clients to controlled tools. Along the way, you will establish the operational foundation that makes those capabilities dependable: generated credentials, persistent storage, health checks, HTTPS options, backups, upgrades, and troubleshooting.

The emphasis is deployment, integration, and informed operation. You will not train a foundation model or compare every available AI framework. Instead, you will learn how a focused set of open source components fits together, how data moves through them, what each service persists, and which decisions remain yours. When a stack supplies infrastructure but not a finished application, the book identifies the client logic or configuration still required.

What self-hosted means here

In this book, self-hosted means that you operate the application services and their persistent data on a host or virtual machine you control. It does not mean that the system is automatically offline, private, secure, or free of external dependencies.

Initial setup can contact source repositories, container registries, and model repositories. A model request can leave your infrastructure if you configure a hosted provider. Calling an MCP tool can send selected data to an external service. Network exposure, logging, backups, account access, and update policy also remain your responsibility.

This distinction shapes every deployment in the book. Self-hosting gives you control over more of the system boundary, but it also makes that boundary yours to understand and operate.

What this book covers

The reference platform combines technologies and maintained containers with distinct roles:

- Ollama runs local large language models (LLMs) through InferCrate.
- LiteLLM routes local and optional hosted model requests through GatewayCrate.
- AnythingLLM provides browser chat, workspaces, and document features.
- Hugging Face Text Embeddings Inference (TEI) converts text into vectors and can rerank candidate results through EmbedCrate.
- PostgreSQL with pgvector stores application and vector data.
- Docling converts documents into structured text or Markdown through ParseCrate.
- Whisper models and faster-whisper power recorded transcription in ScribeCrate, while WhisperLive powers real-time transcription in ScribeCrate Live.
- Kokoro generates speech from text through SpeakCrate.
- MCPHub and Caddy power UplinkCrate's gateway for selected Model Context Protocol (MCP) tools.

The book begins by defining the architecture and choosing a deployment size. Chapter 3 deploys the complete baseline; you can instead start with a smaller stack or one component. Later parts cover model chat, retrieval-augmented generation (RAG), voice, tools, an optional agent, and secure operations.

What you will be able to deploy

By following the path that fits your workload, you can deploy:

- A complete local-first platform in one Docker Compose project
- An API-only local chat stack using InferCrate and GatewayCrate
- Password-protected browser chat using AnythingLLM
- Authenticated document conversion, embeddings, reranking, and vector storage services from which you can build a retrieval pipeline
- Recorded and real-time transcription services, speech output, and a headless multi-stage voice pipeline
- A coding stack with local models, controlled MCP tools, and embeddings for client-built semantic search, or a smaller AI-tools stack without embeddings
- An optional containerized goose agent connected to model and controlled-tool endpoints

These deployments supply services and starting integrations, not every end-to-end application. The lightweight RAG variants still need ingestion and retrieval client logic. The `voice-chat` stack does not automatically connect its speech APIs to browser chat. The relevant chapters identify the remaining client code or UI configuration.

The default platform does not start an agent runtime; Chapter 18 adds goose separately. A compatible agent client or runtime needs its own limits, approval policy, and audit controls.

How the book and companion repositories work together

The maintained Self-Hosted AI Stack repository (github.com/hwdsl2/self-hosted-ai-stack) and its component repositories are the authoritative sources for executable Compose files, container behavior, environment variables, and current commands. This book provides the context those repositories intentionally keep concise: architecture, tradeoffs, request flows, security boundaries, operational practice, and connections among services.

Software changes faster than durable deployment principles, and the companion repositories continue to evolve. If a command or version in the book differs from a maintained repository, review that repository's current release notes, configuration, and instructions before applying the change.

Choose a reading path

You can read in order or choose a deployment in Chapter 2 that fits your workload. Chapter 3 starts the **complete** platform; it is not a required installation step for the other paths:

- **Complete platform:** Deploy in Chapter 3, then use the relevant capability chapters.
- **Private chat:** Start `chat-ui` in Chapter 6 or `chat-only` in Chapter 7. Use Chapters 4 and 5 for model and gateway details.
- **Document intelligence and RAG:** Start a RAG variant in Chapter 11; use Chapters 8 through 10 for its service and database concepts.
- **Speech and voice:** Start a voice variant in Chapter 15; use Chapters 12 through 14 for individual speech services.
- **Coding tools and AI agents:** Start a tool variant in Chapter 17; Chapters 16 and 18 explain gateway permissions and the optional agent.
- **One component:** Start in its component chapter. Plan how it will reach any other service it needs; it does not inherit the complete stack's wiring.
- **Operations:** Read Chapter 3 for platform context, then focus on Part VI. Its deployment block is optional for this reading path.

Dependencies still matter when you skip ahead. Each chapter identifies any service, credential, model, or earlier concept that must already be available. If you read in order but chose a later lightweight start, read the earlier component chapters for concepts and skip their standalone installation blocks. Read Part VI for security, troubleshooting, and recovery on every route.

Choose a deployment size

The complete stack is the clearest way to see how the platform fits together. Its default configuration starts InferCrate, GatewayCrate,

AnythingLLM, PostgreSQL, EmbedCrate, ScribeCrate, and UplinkCrate. ParseCrate, SpeakCrate, and ScribeCrate Live are optional definitions that you enable when needed.

If you need fewer services, use one maintained lightweight variant for chat, retrieval, voice, or tools. Start with the matching chapter above and its companion README. Variants share default names, ports, and volumes: run only one unmodified variant at a time. Chapter 19 covers customization.

If you need one API or already have an application platform, deploy its component independently. For example, standalone InferCrate provides a model API; `chat-only` adds GatewayCrate when you need a gateway. Separate components do not inherit another stack's network or credentials. Chapter 2 compares the paths.

Conventions used in this book

Commands appear without a shell prompt so you can copy them. A command's introduction states the directory from which it should run. Values such as `<server-address>` and `<your-api-key>` are placeholders that you must replace. Never copy angle brackets into a value unless a command explicitly requires them.

Callouts use four labels:

- **Note:** Context that may help but does not affect safety or success
- **Important:** Information required for the procedure to work as intended
- **Security:** A privacy, authentication, authorization, or exposure concern
- **Warning:** A risk of data loss, interruption, unexpected cost, or material system impact

CPU-only deployment is the baseline. NVIDIA CUDA is an optional path where the maintained images support it. Examples use reserved hostnames and addresses, and no example credential is valid.

Before you begin

You should be comfortable using a terminal, editing a text file, and working with basic ports and hostnames. The primary environment is a Linux host with Docker Engine, the Docker Compose v2 plugin, and a non-root user permitted to use Docker. You do not need prior machine-learning, vector-database, GPU, RAG, or MCP experience.

Before buying a GPU, server, or upgraded computer, test a small local model on hardware you already own. If you consider rented cloud capacity, include its cost, network exposure, backups, and administration in the decision.

Companion Resource: Your Local AI Readiness Guide

I created this companion guide to help you get more from this book and make a confident decision before investing more time or money.

Use it to: assess the computer you already own; test a small local model on a representative task; compare on-device AI, self-hosted cloud infrastructure, managed AI, and hybrid approaches; and record your decision and know what would justify revisiting it.

You can download the PDF without registration. Scan the QR code or visit **selfhostedaistack.com/ready**.

Self-hosting gives you more control, but it also transfers operational responsibility to you. Read warnings before destructive commands, keep backups, limit network exposure, and verify results at the application level. A running container is not necessarily a working service.

Chapter 1 begins with the decision that should precede every deployment: which workloads benefit from self-hosting, and which are better left with a managed provider.

Part I: Plan and Deploy Your Platform

Part I turns the decision to self-host into a concrete deployment plan. You will define the boundary of your self-hosted AI system, map services to suitable hardware, and deploy the complete stack with generated credentials and persistent data.

Chapter 1: Why Self-Host AI?

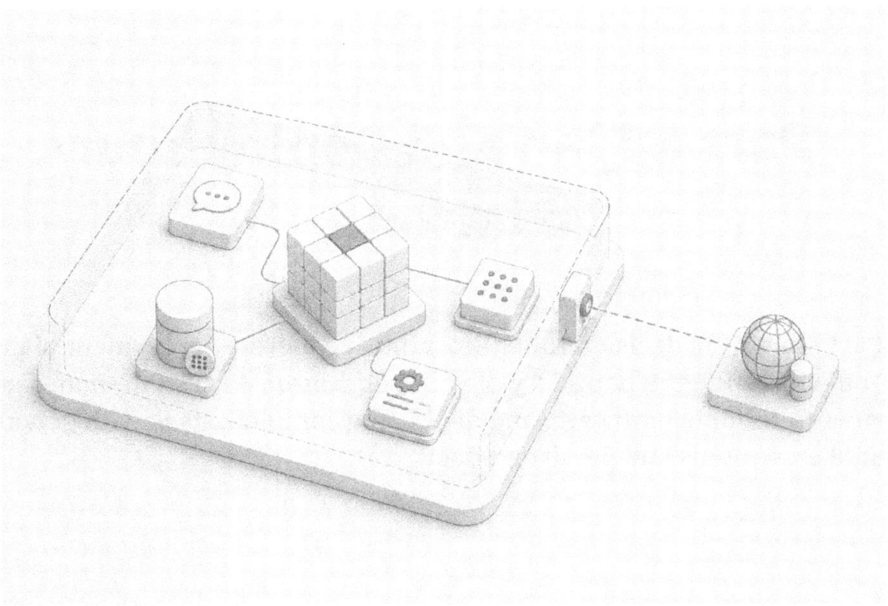

Self-hosting an AI workload means running the service layer on infrastructure you administer. That may be a workstation, a home server, an on-premises host, or a rented virtual or dedicated server. It does not require every model or data source to be local. A self-hosted gateway can route some requests to a local model and others to a hosted provider.

Picture the same model service running on your laptop, on an always-on server in your home or office, or on a rented Linux server. You administer the model service in each case, so each can be self-hosted. A rented server is remote, neither on-device nor available to you offline; its provider still operates the physical host and network. By contrast, when you call a provider-operated model API, that provider runs the model service, even if you operate your own gateway in front of it.

The useful question is not whether self-hosting is always better. It is which parts of your workload benefit from local control and whether you are prepared to operate them.

Chapter 3 deploys this reference implementation after the architecture and hardware planning in Chapter 2.

1.1 Privacy, control, cost, and customization

Self-hosting can provide four practical advantages:

- **Privacy control:** You can keep selected prompts, documents, audio, model requests, and application data inside a boundary you define.
- **Operational control:** You choose when to update, which ports to publish, where data is stored, and how services are backed up.
- **Configuration control:** You can select models, routes, embeddings, tools, and retention behavior without waiting for a hosted product to expose an option.
- **Cost control:** Existing hardware and predictable workloads may reduce per-request provider charges, although hardware, electricity, storage, and maintenance remain real costs.

These are capabilities, not automatic outcomes. Publishing an unauthenticated API, enabling an external provider, invoking a remote search tool, or neglecting backups can defeat the intended benefit. The stack generates credentials and supports restricted bindings, but you must still define and maintain the boundary.

1.1.1 Compare total cost for your workload

Do not compare only a provider's request price with the purchase price of a server. Define the same workload and time period for both options, then record:

Self-hosted costs	Hosted costs
Hardware purchase or rental	Input, output, audio, or tool usage charges
Electricity, cooling, and network service	Minimum commitments or subscriptions
Storage, backups, and replacement capacity	Storage, retrieval, and data-transfer charges
Administrator setup and maintenance time	Integration, review, and vendor-management time
Expected repair and upgrade reserve	Expected growth and price-change reserve

Also record the value of requirements that are not captured by a request price, such as offline availability, data-location constraints, support commitments, or access to a model that does not fit locally. Use a range for uncertain inputs and revisit the comparison after measuring real utilization. There is no universal break-even point.

1.2 Local models versus hosted providers

Local and hosted models solve different constraints:

Consideration	Local model	Hosted provider
Data flow	Can remain on infrastructure you control	Requests reach the provider
Capability	Limited by available models and hardware	Often offers larger or specialized models
Latency	Avoids internet transit but depends on local compute	Depends on network and provider load
Marginal cost	Primarily compute and energy	Usually usage-based pricing
Operations	You manage models, capacity, and availability	Provider manages serving infrastructure
Customization	High control over versions and routing	Limited to exposed provider features

A hybrid design is often practical. GatewayCrate can expose one API while routing local requests to InferCrate and explicitly configured requests to external providers. This gives applications a stable interface without hiding the data boundary. Every external route must still be disclosed and chosen deliberately.

For example, a team may keep confidential notes on an approved self-hosted system and use a hosted model for public drafts when policy permits. Check the whole request: earlier messages, attached text, and retrieved passages may also travel with the question.

Model quality is workload-specific. A small local model may be adequate for classification, summarization, private drafts, or tool selection, while another task may justify a hosted model. Test with representative inputs rather than choosing by parameter count or marketing claims alone.

1.3 What self-hosting does and does not solve

Self-hosting can reduce dependency on a single hosted application and give you direct control over configuration and stored data. It does not by itself solve:

- Weak authentication or unnecessarily published ports
- Vulnerabilities in the host, container images, or upstream applications
- Prompt injection or unsafe MCP tool permissions
- Inaccurate, biased, or fabricated model output
- Capacity planning, monitoring, backups, and recovery
- Licensing or policy obligations for models and data
- Data sent through external providers or remote tools you enable

The host, network, container runtime, applications, and model integrations form a shared system. Security and privacy depend on the complete request path, not the location of one model.

1.4 Common use cases

The platform in this book supports several common patterns:

- **Private chat:** AnythingLLM sends browser chat through GatewayCrate to a local model served by Ollama.
- **Document retrieval:** ParseCrate prepares files, EmbedCrate converts chunks to vectors, pgvector stores them, and client logic retrieves context for a model.
- **Transcription:** ScribeCrate transcribes recordings. ScribeCrate Live handles live audio.
- **Speech output:** SpeakCrate generates audio from model or application text.
- **Coding assistance:** A compatible client uses GatewayCrate for models, EmbedCrate for semantic lookup, and UplinkCrate for selected tools.
- **AI agent integration:** An external runtime uses the platform's model and tool endpoints under explicit limits.

These services are building blocks. A deployed API is not automatically a complete workflow, and a tool-capable client is not necessarily an autonomous agent.

1.5 Decide whether this platform fits your needs

Self-hosting is a strong candidate when:

- You need control over data location, retention, models, or updates.
- Your team can administer a Linux host and containerized services.
- The workload fits hardware you have or can operate economically.
- You value local availability or experimentation more than a managed service guarantee.
- You can test, patch, monitor, back up, and recover the deployment.

A managed service may be the better starting point when:

- You need a capability that available local models cannot provide.
- You cannot maintain the host or respond to security updates.
- Workload demand is highly variable and idle hardware would dominate cost.
- Formal support, availability commitments, or certified controls are required.
- Sending the relevant data to an approved provider is acceptable.

A hybrid design is appropriate when tasks differ in sensitivity, quality, or capacity requirements. Document the rule that selects each route so an application that appears local does not silently send sensitive content elsewhere.

1.6 Summary and next steps

Self-hosting is a control decision with operational consequences. It can keep selected workloads local and make models, storage, and tools configurable, but it does not remove the need for security, testing, or clear data boundaries.

Chapter 2 turns that decision into an architecture and hardware plan. You will choose the service set, deployment location, CPU or GPU path, model scale, and storage capacity needed before creating containers.

Chapter 2: Choose Your Architecture and Hardware

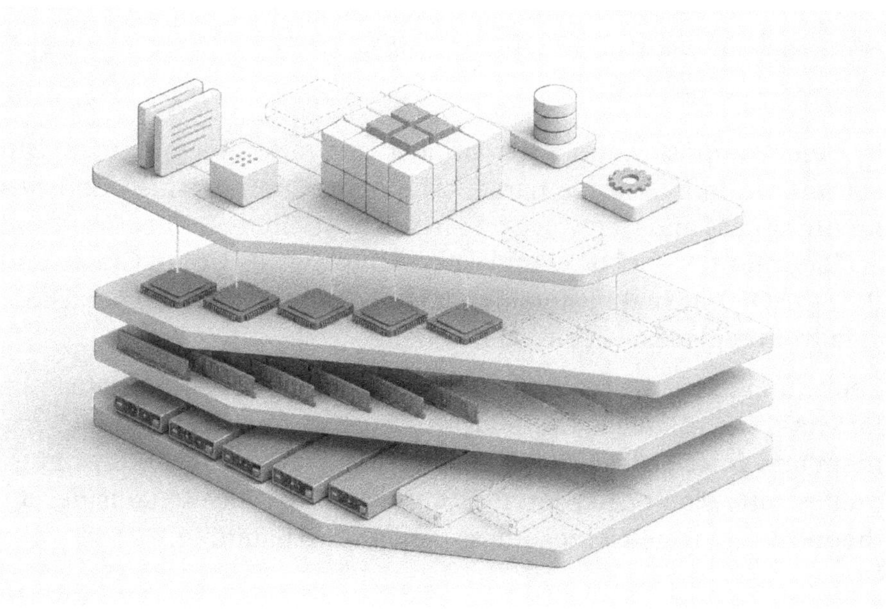

Before deploying, choose where the platform will run, which services it needs, and how demanding its model and application workloads will be. The goal is not a universal hardware recommendation. It is a deployment profile that makes your assumptions explicit and can evolve with the workload.

The reference platform includes the service compositions, CPU and CUDA variants, architecture support, and current baseline requirements used in this chapter. Choose a route here before starting any Compose project. Chapter 3 deploys the complete platform; smaller routes begin in later chapters.

2.1 Understand the platform architecture

The reference platform separates user interfaces, gateways, model servers, specialized AI APIs, tools, and storage:

```
Users and compatible clients
      |
      +--> AnythingLLM --> LiteLLM --> Ollama
      |                      |
      |                      +----> optional external
 providers
      |                      +----> MCP gateway
      |
      +--> Whisper or WhisperLive
      +--> Docling --> TEI --> pgvector
      +--> Kokoro
```

Docker Compose supplies a project network so containers can communicate by service name. The main Compose file publishes AnythingLLM and GatewayCrate to the host, binds EmbedCrate and ScribeCrate to loopback, and keeps InferCrate, UplinkCrate, and PostgreSQL internal by default. Named volumes separate persistent data from replaceable containers.

The automatically wired path is AnythingLLM to GatewayCrate to InferCrate, with GatewayCrate also connected to UplinkCrate. Document and voice services expose APIs that a client or workflow must orchestrate. Treat each service boundary as a place to consider authentication, data flow, readiness, and failure handling.

2.2 Estimate CPU, memory, storage, and GPU requirements

The complete stack requires at least 8 GB of memory with a small model. This is a startup floor, not a performance promise. Lightweight variants start at about 4.5 GB, with the actual requirement determined by the selected model and active services.

Make a provisional **go, test, or pause** decision before buying hardware. For example, suppose an existing host has 8 GB RAM, with 2 GB already used by the operating system and other applications. About 6 GB remains for Docker, the model, and request peaks. The maintained `chat-only` estimate is about 4.5 GB with `llama3.2:3b` and a short context, leaving about 1.5 GB of apparent headroom. The task is one active request to summarize a synthetic 500-word note into three accurate bullets within 60 seconds. These are illustrative planning figures, not measurements or a performance guarantee. The decision is **test**: try the task with one local model in the trial below. If it is useful, test the chosen Compose variant and measure the whole host before relying on the 4.5 GB estimate. A larger model, long context, or second active request could consume the margin. If this workload cannot fit with safe headroom, pause or choose a smaller service set before buying a server.

Estimate capacity from these workload drivers:

- **Model weights and quantization:** Larger or higher-precision weights take more storage and require more RAM or video memory (VRAM) when loaded.
- **Context length and KV cache:** During inference, the runtime stores key and value tensors for tokens in the active context in a key-value cache, or KV cache. Longer active contexts generally enlarge this cache.
- **Concurrency:** Parallel model sequences can require separate KV cache capacity. Simultaneous embedding, transcription, or document requests also add their own working memory and compute demand.
- **Specialized services:** Document conversion, diarization, reranking, and voice services can load additional models.
- **Persistent data:** Downloaded models, caches, chat data, documents, database files, generated audio, and backups consume storage over time.

Measure steady-state and peak usage with representative requests. Leave margin for the operating system, Docker, updates, temporary files, and recovery operations. A host that barely starts the stack has no useful failure margin.

Use the deployment profile worksheet in Appendix B to record the estimate. Keep measured values separate from assumptions. A useful first profile records idle memory, peak memory during one representative request, model storage, request latency, and the highest tested concurrency. Measurements from another machine are useful reference points, not capacity guarantees for yours.

2.3 Choose CPU-only or NVIDIA CUDA deployment

The CPU Compose files provide the broadest baseline and support the maintained multi-architecture deployment path. They require no GPU runtime and work well for functional testing, small models, and lower-volume APIs.

The CUDA files select GPU-enabled images for supported compute-heavy services and request NVIDIA devices through Docker. In the complete stack, InferCrate and ScribeCrate use CUDA images by default. Optional ParseCrate, SpeakCrate, and ScribeCrate Live definitions also contain GPU reservations. CUDA images are limited to `linux/amd64` in the current sources and require a compatible NVIDIA driver and NVIDIA Container Toolkit.

Choose CPU first when compatibility, simplicity, or low request volume matters more than latency. Choose CUDA when tested workloads justify the additional hardware, driver, and runtime dependencies. A GPU can reduce inference time, but it does not accelerate every service or remove CPU and system-memory needs.

2.4 Select models that fit your hardware

Model names and rankings change quickly, but the selection method is durable:

1. Define the task and a representative evaluation set.
2. Determine the RAM or VRAM available after reserving host and service margin.
3. Choose the smallest model likely to meet the task.
4. Test quality, latency, context use, and concurrent behavior.
5. Increase model size only when the measured benefit justifies the resources.

Quantization can reduce model-weight storage and loaded-weight memory, with quality and performance tradeoffs that depend on the model, runtime, hardware, and workload. A model file size or weights-only estimate is not total runtime memory. Add the KV cache, runtime buffers and overhead, other loaded models, active services, and host margin. Context limits advertised by a model do not guarantee that the host can serve that context efficiently, especially with parallel requests.

The stack's starter example uses `llama3.2:3b` because a small model makes the deployment path accessible. It is an installation check, not a recommendation for every workload. Chapter 4 develops the model-selection and verification process further.

2.5 Choose a local server, workstation, or cloud host

Location	Advantages	Main tradeoffs
Workstation	Existing hardware, direct access, useful for development	Competes with interactive work and may not run continuously
Local server	Stable local service and storage boundary	Hardware, power, cooling, and remote-access responsibility
On-premises host	Organizational control and internal-network integration	Procurement and operations overhead
Cloud host	Fast provisioning and accessible capacity	Provider trust, recurring cost, egress, and public-network exposure

DigitalOcean, Vultr, and OVHcloud are examples of providers with rented compute options. Their product names, available capacity, and prices change; the choice here is whether you will operate the model service on a rented host or use a provider-operated model API. A remote CPU virtual machine does not automatically make a demanding model fast.

The physical location alone does not define the trust boundary. Consider host administrators, hypervisor or cloud operators, backups, DNS, reverse proxies, external model providers, remote MCP tools, and the clients that hold API keys.

For an internet-facing deployment, plan DNS, TLS, firewall rules, and recovery before publishing a port. For a local-only deployment, decide whether clients will use the host itself, a trusted LAN, a VPN, or an SSH tunnel.

2.5.1 Try one model before choosing a host

If your computer seems too small, test one limited task before deciding whether to keep using it, rent a server, or pause. The companion readiness guide at **selfhostedaistack.com/ready** walks through installing and running one small local model with Ollama on your computer. Use invented, nonsensitive text. For example, write a synthetic 500-word note, ask for three accurate summary bullets, and time the answer. Note whether the computer's ordinary applications remain usable. Use the guide for installation steps and the model name; this trial needs no Docker stack.

If that task works, a small personal workload may fit your computer, but the result does not establish that the complete platform or multiple users will fit. If it fails, you have learned about this machine, model, and task, not every self-hosted route. You can try a smaller workload, plan an owned or rented server, choose an approved managed service, or pause. Before renting GPU or dedicated capacity, set a cost limit and plan access control, permitted data, backups, and recovery. The guide compares these paths in more detail.

2.6 Choose the deployment path before you install

Select one starting point by the work you need now. Component chapters explain services already running in your chosen stack; their independent installation blocks are alternatives, not steps to run in sequence.

Starting path	Choose it when	Start in
Complete platform	Several capabilities justify the full service set	Chapter 3
One lightweight variant	A defined chat, retrieval, voice, or tool workload needs fewer services	Chapter 6, 7, 11, 15, or 17
Standalone component	One API is enough, or another platform supplies its dependencies	The component chapter

If a lightweight variant fits, choose one by the interface and services your workload needs:

Need	Starting stack
Model and gateway APIs without browser chat	chat-only
Password-protected browser chat	chat-ui
Embeddings for a retrieval application	rag-pipeline
Embeddings plus document conversion	rag-pipeline-full
Recorded speech and speech output APIs	voice-pipeline
Browser chat plus separate voice APIs	voice-chat
Models and controlled MCP tools	ai-tools
Models, controlled MCP tools, and embeddings	code-assistant

Every lightweight variant starts InferCrate, GatewayCrate, and PostgreSQL. In Chapter 11, `rag-pipeline` adds EmbedCrate and `rag-pipeline-full` also adds ParseCrate. In Chapter 15, `voice-pipeline` adds ScribeCrate and SpeakCrate; `voice-chat` also adds AnythingLLM. In Chapter 17, `ai-tools` adds UplinkCrate and `code-assistant` also adds EmbedCrate. Their documented 3B-model estimates range from about 4.5 to 6.5 GB, depending on the services; measure your chosen workload.

For a model API alone, standalone InferCrate in Chapter 4 can suffice. Choose `chat-only` when the client needs GatewayCrate's gateway and its PostgreSQL-backed configuration (Chapter 7). For password-protected browser chat, choose `chat-ui` (Chapter 6). For a document, speech, and tool platform, the complete stack may be simpler to operate than several independently connected containers. It still leaves ParseCrate, SpeakCrate, and ScribeCrate Live disabled until you enable them. A RAG or voice stack provides APIs, not a finished ingestion client or browser voice flow.

The maintained lightweight variants reuse default names, ports, and volumes. Do not run two unmodified variants or a variant and the complete stack side by side. If one variant lacks a service you need, choose another maintained variant, enable an optional definition where provided, or plan the networking and credentials for a separate component. Chapter 19 covers deliberate combination.

Consider how these choices play out. **One person, one task:** Suppose the local trial above gives accurate summaries quickly enough and the computer stays usable. That reader can keep the native Ollama setup for now; there is no need to run the complete platform. If a separate application later needs a model-and-gateway API, `chat-only` in Chapter 7 is a smaller next step. If the trial is too slow, the reader should test the same task on a candidate host before committing to more capacity. A remote CPU server can offer an always-on location, but it does not guarantee faster or better answers.

A small team, shared chat: Suppose three colleagues need an always-on, password-protected browser for permitted internal notes, but no speech or tool services. A workstation that sleeps is a poor shared host. They could evaluate a rented Linux server from Vultr as one example, or use an owned server. First confirm that their data policy permits the provider, that the selected model and expected simultaneous use fit the host, and that someone will manage access, updates, backups, and recovery. A CPU virtual machine may keep the service available without making inference fast; a GPU rental needs a bounded cost and a representative test before any commitment. If those conditions are met, they can start with `chat-ui` in Chapter 6, restrict direct port access, and plan HTTPS for remote users as Chapter 20 describes. Their first check is that colleagues can sign in and obtain a useful answer from the model on that host within an acceptable time. If provider trust is the obstacle, consider an owned host. If operating a server is the obstacle, an approved managed service may fit, but model requests then cross a different data boundary. The complete platform is justified when its additional services serve an actual workload.

A team with several workflows: Now suppose the same colleagues also need to transcribe permitted recordings and build a separate document search application with per-user access rules. This goes beyond asking questions about uploaded documents in AnythingLLM. They have a developer for that application and someone to maintain the host. They could connect separate deployments, but each connection adds network paths, credentials, and data to back up. The complete platform in Chapter 3 is a reasonable starting point: its default Compose project includes chat, local model routing, embeddings, recorded transcription, and PostgreSQL with pgvector support. It also starts UplinkCrate, even if they have no approved tool use yet. ParseCrate, SpeakCrate, and ScribeCrate Live remain disabled until the team enables them for a defined task.

The larger stack still does not implement their search application or connect ScribeCrate's transcriptions to browser chat. The developer must build and verify ingestion and retrieval in a client, control which documents each user may retrieve, and arrange any transcription workflow the team wants. Before deploying, they should test representative requests and simultaneous use on the chosen host, and assign responsibility for access, backups, and recovery. If they need only shared chat and workspace document questions, `chat-ui` remains the smaller choice. If the separate workflows are real and the team can operate them, proceed to Chapter 3 and then use the relevant later chapters to build and verify each workflow.

2.7 Plan storage and backups

Separate storage planning into categories:

- **Regenerable but expensive:** Container images and downloaded model caches
- **Configuration and credentials:** Generated API keys, database secrets, and service settings
- **User and application data:** Chats, workspaces, uploaded documents, vector records, and database state
- **Generated outputs:** Transcripts, converted documents, and audio retained outside API responses
- **Recovery copies:** Backups, checksums, and enough temporary capacity to restore safely

The first category can usually be downloaded again. The others may be unique. Place Docker's data root and backup destination on storage appropriate to their importance. A second directory on the same disk does not protect you from that disk failing.

Chapter 3 identifies the root stack's volumes. Chapter 22 provides backup, restore, migration, and upgrade procedures.

2.8 Planning checklist

Record a deployment profile before continuing:

- ☐ Host location, administrator, and trust boundary
- ☐ Linux distribution and CPU architecture
- ☐ CPU-only or NVIDIA CUDA path
- ☐ Available RAM, VRAM, storage, and expected growth

- ☐ Starter model and representative workload
- ☐ Required services and optional services left disabled
- ☐ Expected concurrent users or requests
- ☐ Client access path and required published ports
- ☐ External providers or tools and their permitted data
- ☐ Backup destination and recovery owner
- ☐ Update and maintenance window

Appendix B provides a reusable deployment profile worksheet based on this checklist. Update it after the first deployment and after any model, service, or concurrency change that materially affects capacity.

2.9 Summary and next steps

You now have the assumptions needed to deploy deliberately: a host, service set, resource envelope, model starting point, access boundary, and recovery plan. Preserve that profile so later performance or security changes can be compared with the original design.

If you chose the complete platform, Chapter 3 applies the plan and verifies the first local chat request. Otherwise, start at the chapter named for your route above.

Chapter 3: Deploy the Complete Self-Hosted AI Stack

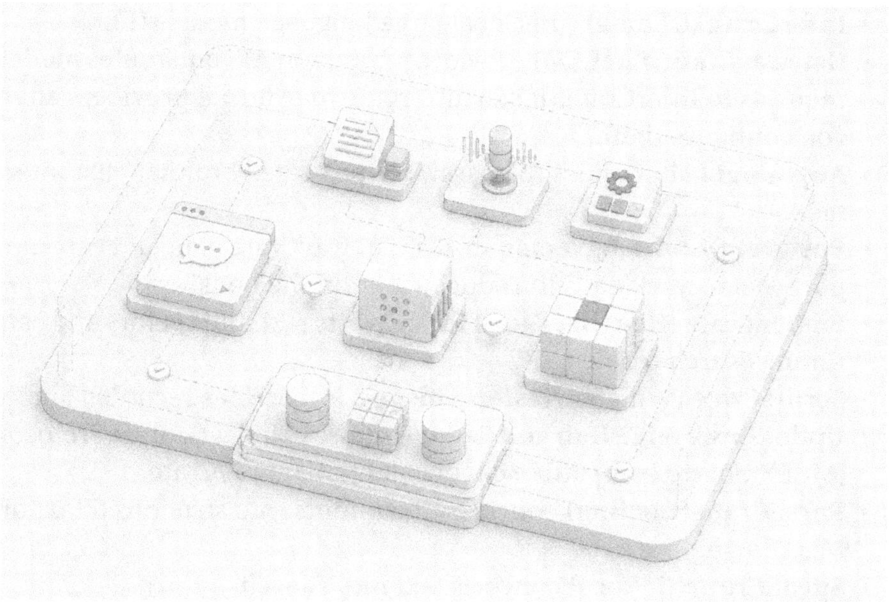

By the end of this chapter, you will have a working self-hosted AI platform with a browser-based chat interface, local model inference, a model gateway, speech transcription, embeddings, a tool gateway, and persistent storage. You will also know which services are reachable from the host, where their data is kept, how to verify the request path, and how to stop the platform without deleting its data.

The repository's Docker Compose files encode the supported service relationships, generated secrets, health checks, volumes, and optional NVIDIA CUDA configuration. This chapter explains those decisions and provides checkpoints around the concise procedure. Later chapters examine each component and workflow in more depth.

Follow this deployment procedure only if you selected the complete platform in Chapter 2. A lightweight or standalone reader can use its architecture and security explanations without starting this Compose project. In later chapters, reuse services already running here instead of repeating their independent installation steps. The 8 GB minimum below applies to this complete startup, not to every route in the book.

3.1 Understand the included services

The stack is a platform of cooperating services, not one monolithic AI application. Each component has a focused responsibility:

- **InferCrate (Ollama)** runs local large language models (LLMs).
- **GatewayCrate (LiteLLM)** provides a common API and routes model requests to InferCrate. It can also route to external providers after you configure them.
- **AnythingLLM** provides the password-protected browser chat interface.
- **PostgreSQL with pgvector** stores GatewayCrate data and can support vector workloads introduced later in the book.
- **EmbedCrate (Hugging Face TEI)** converts text into vectors and can rerank search results.
- **ScribeCrate (Whisper/faster-whisper)** transcribes recorded audio.
- **UplinkCrate (MCPHub and Caddy)** exposes Model Context Protocol (MCP) tools to compatible clients and to GatewayCrate.
- **ParseCrate (Docling)** converts documents into structured text or Markdown.
- **SpeakCrate (Kokoro)** converts text into speech.
- **ScribeCrate Live (WhisperLive)** transcribes live audio streams.

The default deployment starts InferCrate, GatewayCrate, AnythingLLM, PostgreSQL, EmbedCrate, ScribeCrate, and UplinkCrate. ParseCrate, SpeakCrate, and ScribeCrate Live are present in the Compose files but commented out to reduce initial memory use. You can enable them when the relevant chapters introduce their workflows.

These boundaries matter. The stack automatically connects AnythingLLM to GatewayCrate, GatewayCrate to InferCrate, and GatewayCrate to UplinkCrate. The EmbedCrate and ScribeCrate services start as authenticated APIs, but their presence does not by itself create a finished retrieval-augmented generation (RAG) or voice application. Those workflows require the additional choices and request logic covered in Parts III and IV.

The main local chat request follows this path, labeled with the underlying technologies and the maintained containers' ports:

```
User's browser
       |
       | HTTP :3001
       v
AnythingLLM  ----->  LiteLLM  ----->  Ollama
                       |            local model
                       +--------->  MCP gateway
                       |
                       +--------->  PostgreSQL

Separate local APIs: TEI :8000 and Whisper :9000
Optional local APIs: Docling :5001, Kokoro :8880,
                     WhisperLive :9090 and :8001
```

Docker creates a project-scoped bridge network. Containers communicate by service name. For example, GatewayCrate reaches InferCrate at `http://ollama:11434` without publishing the InferCrate port on the host. Shared volumes carry generated InferCrate and UplinkCrate keys to GatewayCrate, so you do not have to paste those secrets into configuration files.

The default host exposure is intentionally uneven:

Service	Default host access
AnythingLLM	Port 3001 on all host interfaces
GatewayCrate	Port 4000 on all host interfaces
EmbedCrate	127.0.0.1:8000 only
ScribeCrate	127.0.0.1:9000 only
InferCrate, UplinkCrate, PostgreSQL	Docker network only
Optional ParseCrate, SpeakCrate, ScribeCrate Live	Loopback only when enabled

Security: AnythingLLM and GatewayCrate use plain HTTP in the baseline deployment. Their generated credentials prevent anonymous use, but HTTP does not encrypt those credentials in transit. Limit ports `3001` and `4000` to the host or a trusted network during this chapter. Do not expose them directly to the internet. Chapter 20, "Secure Local and Internet-Facing Services," adds localhost binding, HTTPS, and the Caddy reverse-proxy overlay.

A local chat path does not guarantee that every possible request remains on the host. The initial deployment downloads code, images, and models. Configuring an external provider in GatewayCrate can send prompts and responses to that provider. MCP tools such as `fetch` can contact external destinations when a client invokes them. Keep the baseline test local, and review each added integration's data flow before using it with sensitive information.

3.2 Prepare Docker and Docker Compose

Start with the CPU-only configuration, even if the host has a GPU. This keeps general deployment problems separate from GPU runtime problems. Later in the chapter, you can switch the same persistent deployment to CUDA.

Prepare a Linux host with:

- Docker Engine and the Docker Compose v2 plugin
- Git for retrieving the maintained Compose files
- A non-root user that has permission to use Docker
- At least 8 GB of memory for the complete stack with a small local model
- Enough free storage for container images, model files, application data, and future backups
- Outbound network access for the repository, container images, and initial model downloads

Larger models, longer contexts, and concurrent requests need more memory than the baseline. Chapter 2 explains those variables. The stack quick-start documentation is the authority for supported host architectures and changing prerequisites.

Confirm that the required commands work for the current user:

```
git --version
docker --version
docker compose version
docker info >/dev/null
```

Each command should exit successfully. If `docker info` reports a permission error, correct the user's Docker access before continuing. Do not alternate between privileged and unprivileged Docker commands because that can create confusing differences in visible containers and files.

Review memory and filesystem capacity before pulling images and models:

```
free -h
df -h
```

Also confirm that the host does not already use the main published ports. The exact command depends on the Linux distribution, but a typical check is:

```
ss -ltn | grep -E ':(3001|4000|8000|9000)[[:space:]]' || true
```

No output means those four ports are not currently listening. Output is not automatically an error, but you must identify the existing listener before the deployment. Chapter 19 covers supported port customization.

3.3 Deploy the stack

Choose a directory for maintained application checkouts. From its parent directory, clone the repository and enter it. If you already have the repository, reuse that checkout and begin with `cd self-hosted-ai-stack`.

```
git clone https://github.com/hwdsl2/self-hosted-ai-stack
cd self-hosted-ai-stack
```

All remaining commands in this chapter run from the repository root unless a step says otherwise.

Validate the CPU Compose file before creating containers:

```
docker compose config --quiet
```

No output and a zero exit status indicate that Docker Compose rendered the configuration successfully. This check catches syntax and interpolation problems, but it does not prove that images can be downloaded or that services will become ready.

Start the stack with the standard command:

```
docker compose up -d
```

Docker Compose creates the network and named volumes, pulls missing images, and starts services in dependency order. The one-shot `ai-stack-init` container initializes the PostgreSQL secret before the database starts. InferCrate and UplinkCrate generate their API keys, GatewayCrate reads the shared copies, and the AnythingLLM bootstrap creates its initial administrator password and persistent environment file.

A successful return from the Compose command does not mean every application is ready. PostgreSQL must initialize, GatewayCrate waits for its dependencies, and AnythingLLM performs its own startup and database work. Inspect the current state:

```
docker compose ps --all
```

The long-running services should progress toward `Up` or `healthy` states. An `ai-stack-init` container that exits with status `0` has completed normally. An exit status other than `0` requires investigation.

InferCrate starts without a language model. Pull the small starter model used by the current stack defaults:

```
docker exec ollama ollama_manage --pull llama3.2:3b
```

The download is stored in `ollama-data`, so container recreation does not repeat it. Model availability and naming change more quickly than the platform architecture. Check the current stack instructions if this example no longer resolves, and use Chapter 4 to choose a model appropriate for your hardware and workload.

3.4 Retrieve generated credentials

Fresh persistent deployments generate unique credentials rather than relying on passwords printed in the Compose file. Retrieve credentials only after the corresponding container is running.

Display the three core API keys and their endpoint information:

```
docker exec ollama ollama_manage --showkey
docker exec litellm litellm_manage --showkey
docker exec mcp mcp_manage --showkey
```

The default deployment also generates keys for EmbedCrate and ScribeCrate:

```
docker exec embeddings embed_manage --showkey
docker exec whisper whisper_manage --showkey
```

For AnythingLLM, retrieve the initial administrator password from its persistent volume:

```
docker exec anythingllm \
  cat /app/server/storage/.initial_admin_password
```

You can also find it in the first-start log while that log entry remains available:

```
docker compose logs anythingllm | grep -A4 'FIRST RUN'
```

Open `http://<server-address>:3001`, sign in with the generated password, and change it in the security settings. Replace `<server-address>` with `localhost` when the browser runs on the Docker host, or with the server's trusted-network address when accessing it remotely.

Important: The `.initial_admin_password` file records the seeded password. It is not updated when you change the password in AnythingLLM, so it may no longer represent the current login credential.

GatewayCrate's LiteLLM administration interface is at `http://<server-address>:4000/ui`. Sign in with username `admin` and the GatewayCrate master key as the password.

The initialization container also generates a PostgreSQL password and stores it in `ai-stack-shared`. PostgreSQL and GatewayCrate read that file directly. You do not need to display, copy, or place this database password in a shell command for the standard deployment. Existing deployments that used a custom database password need the compatibility procedure in the stack's PostgreSQL credential documentation.

Security: Treat all displayed values as secrets. Do not paste them into issues, screenshots, chat sessions, version control, or shell commands that will be shared. A generated API key authenticates a request, but it does not encrypt plain HTTP traffic.

3.5 Understand persistent storage

Containers are replaceable. The named Docker volumes hold the state that must survive a container update or recreation.

Volume	Main contents
ollama-data	Local models, InferCrate configuration, and API key
litellm-data	GatewayCrate configuration and master key
litellm-db	PostgreSQL database files
ai-stack-shared	PostgreSQL secret and stack initialization state
anythingllm-data	Chat data, workspaces, files, password state, and settings
mcp-data	UplinkCrate keys and configuration
embeddings-data	Embedding model cache, settings, and API key
whisper-data	ScribeCrate model cache, settings, and API key

Enabling ParseCrate, SpeakCrate, or ScribeCrate Live adds `docling-data`, `kokoro-data`, or `whisper-live-data`, respectively. The `ollama-shared`, `mcp-shared`, and `litellm-shared` volumes exchange credentials between containers. They are reconstructable coordination data rather than the primary backup copies.

List the named volumes on the host:

```
docker volume ls --format '{{.Name}}' | sort
```

These volume names are explicit and are not prefixed with the Compose project name. This makes them stable across the full and lightweight stack variants, but it also means those variants share default data and container names. Run only one default variant at a time.

Important: Persistent does not mean backed up. A disk failure, an accidental volume deletion, or a damaged application database can still destroy data. Chapter 22, "Back Up, Restore, Upgrade, and Migrate," provides the tested backup and recovery procedure. Back up before upgrades or material configuration changes.

3.6 Verify the installation

Verification has three layers: container state, service readiness, and a useful end-to-end request. Passing only the first layer is not enough.

First, inspect all container states and recent initialization logs:

```
docker compose ps --all
docker compose logs --tail=50 ai-stack-init
```

The initialization log should report that the PostgreSQL secret was generated, reused, or initialized from an explicit override. It must not report an error.

Next, run the repository's health-check script (stack-check.sh) from the repository root:

```
./stack-check.sh
```

The script discovers running services and verifies the checks that apply to them. For the baseline stack, this includes:

- At least one Ollama model and an InferCrate API key
- PostgreSQL readiness and a readable password secret
- GatewayCrate readiness, generated key, database integration, and model routing
- Authenticated EmbedCrate and ScribeCrate endpoints
- An authenticated MCP initialization handshake
- AnythingLLM's health endpoint

Services that remain disabled are reported as skipped. Warnings do not cause a nonzero exit status, but you should read and resolve unexpected warnings before considering the platform ready. A failed check returns a nonzero status. On a first start, wait a few minutes and rerun the check before changing configuration.

You can also make a direct request through GatewayCrate. Store the master key in a temporary shell variable, send a nonstreaming chat request, and remove the variable afterward:

```
gateway_master_key="$(docker exec litellm litellm_manage --
  getkey)"

curl -sS http://localhost:4000/v1/chat/completions \
  -H "Authorization: Bearer $gateway_master_key" \
```

```
  -H 'Content-Type: application/json' \
  -d '{
    "model": "ollama/llama3.2:3b",
    "messages": [
      {"role": "user", "content": "Reply with: stack ready"}
    ],
    "max_tokens": 20,
    "stream": false
  }'

unset gateway_master_key
```

A successful response is JSON with an assistant message under `choices` . The wording may differ from the requested phrase because model output is nondeterministic. An authentication error points to the GatewayCrate key. A model error usually means that the model was not downloaded or that the requested name does not match the configured model.

Finally, sign in to AnythingLLM and keep the stack's default local model route. Paste the synthetic note you used in Chapter 2, or create a new nonsensitive one, and ask for three accurate summary bullets. Check that you can sign in, receive an answer through the browser, and verify its facts against the note. Record how long it takes. With that route, a response verifies the browser, AnythingLLM, GatewayCrate, and InferCrate path together; a useful response also tests your first workload. If the route works but the answer is poor or too slow, use Chapter 4 to test a model and settings that fit your hardware. Do not add an external provider during this baseline test. Keeping the first request local makes the trust boundary and any failure easier to understand.

3.7 Enable NVIDIA CUDA acceleration

The CUDA Compose file uses GPU-enabled images and requests one NVIDIA GPU for InferCrate and ScribeCrate. GPU blocks are also present for the optional ParseCrate, SpeakCrate, and ScribeCrate Live services when you enable them. GatewayCrate, AnythingLLM, UplinkCrate, EmbedCrate, and PostgreSQL continue to use their standard images in this configuration.

Before switching, confirm all of the following:

- The host uses the `linux/amd64` architecture supported by the CUDA images.
- A compatible NVIDIA GPU and driver are installed.
- `nvidia-smi` works on the host.
- NVIDIA Container Toolkit is installed and configured for Docker.
- The current minimum driver requirements in the stack's CUDA instructions are met.

From the repository root, render the CUDA configuration before applying it:

```
docker compose -f docker-compose.cuda.yml config --quiet
```

Then reconcile the existing deployment with the CUDA file:

```
docker compose -f docker-compose.cuda.yml up -d
```

The command replaces containers whose image or device configuration changed. It reuses the same named volumes, so downloaded models, generated keys, chat data, and database state remain in place.

Use the same Compose file for subsequent CUDA management commands:

```
docker compose -f docker-compose.cuda.yml ps
docker compose -f docker-compose.cuda.yml logs --tail=100
  ollama whisper
./stack-check.sh
```

The health check confirms that the services and routing work, not that a model actually used the GPU. While an inference or transcription request is running, use `nvidia-smi` on the host and inspect the relevant service logs to confirm GPU activity.

If you prefer shorter commands for the current shell session, set the Compose file once:

```
export COMPOSE_FILE=docker-compose.cuda.yml
docker compose ps
```

Run `unset COMPOSE_FILE` before returning to the CPU configuration. Chapter 19 covers persistent Compose selection and Podman's different GPU mechanism.

3.8 Stop, restart, and remove the stack safely

Use the least disruptive lifecycle command that accomplishes the task.

Stop the running containers without removing them:

```
docker compose stop
```

Resume the stopped deployment while reapplying dependency checks:

```
docker compose up -d
```

Restart a particular long-running service without changing its configuration:

```
docker compose restart litellm
```

After editing a Compose file or changing an image, use `up` so Docker Compose can reconcile the configuration:

```
docker compose up -d
```

Remove the containers and the default network while preserving named volumes:

```
docker compose down
```

Running `docker compose up -d` later recreates the containers and reconnects the existing volumes.

Warning: Do not add `--volumes` or `-v` to `docker compose down` unless you intend to delete persistent stack data and have verified backups. Volume removal can erase models, API keys, database records, chat history, uploaded documents, application settings, and optional service caches.

When managing a CUDA deployment without `COMPOSE_FILE`, include `-f docker-compose.cuda.yml` in each Compose command. Mixing command contexts can make the displayed configuration differ from the one you intended to run.

3.9 Common first-deployment problems

Start diagnosis with state, the repository health-check script (stack-check.sh), and focused logs:

```
docker compose ps --all
./stack-check.sh
docker compose logs --tail=100 litellm
```

The example inspects GatewayCrate. Substitute another Compose service name such as `ollama`, `db`, `mcp`, or `anythingllm` when a different check fails. Redact keys, passwords, provider credentials, private URLs, and sensitive user data before sharing output.

3.9.1 A service is still starting

First-start model downloads, database initialization, and AnythingLLM setup can take several minutes. Wait, rerun `./stack-check.sh`, and inspect the affected service's logs. The application inside a running container may still be unable to accept requests.

3.9.2 The initialization container exited

An `ai-stack-init` exit status of `0` is expected because it is a one-shot helper. For any other status, inspect its log:

```
docker compose logs ai-stack-init
```

Do not delete `litellm-db` or `ai-stack-shared` to force a retry. Those volumes contain database state and its corresponding secret.

3.9.3 The health check reports no Ollama models

List the downloaded models, then repeat the starter-model pull if necessary:

```
docker exec ollama ollama_manage --listmodels
docker exec ollama ollama_manage --pull llama3.2:3b
```

If the pull fails, inspect available storage, outbound network access, and the InferCrate logs.

3.9.4 A published port is already in use

Docker reports a bind error when another process owns ports `3001`, `4000`, `8000`, or `9000`. Identify the listener instead of stopping an unknown process blindly. Then either remove the conflict or apply the deliberate port changes described in Chapter 19.

3.9.5 AnythingLLM does not accept the initial password

Confirm that the container is ready, retrieve the password from the persistent file again, and check whether someone already changed it in the interface. The initial password file does not follow later password changes.

3.9.6 The host runs out of memory

Stop the stack before the operating system repeatedly kills containers. Confirm which services and models are consuming memory. Select a smaller model, disable unneeded services, or use a lightweight stack rather than adding restart loops or swap as an unexplained workaround.

3.9.7 CUDA services start without using the GPU

Confirm that you used `docker-compose.cuda.yml`, that `nvidia-smi` works on the host, and that NVIDIA Container Toolkit is configured. Review the InferCrate and ScribeCrate logs. Chapter 21 provides a fuller GPU diagnostic sequence.

For persistent startup, routing, networking, or credential problems, continue with Chapter 21, "Diagnose Problems and Verify Stack Health." Preserve the current volumes while diagnosing unless a documented recovery procedure specifically requires replacement.

3.10 Summary and next steps

If you followed the complete deployment, you now have a persistent platform with a verified local chat path from AnythingLLM through GatewayCrate to InferCrate. PostgreSQL, EmbedCrate, ScribeCrate, and UplinkCrate provide the database, retrieval, transcription, and tool foundations used later in the book. Optional document conversion, speech synthesis, and real-time transcription services remain available without consuming resources until you enable them.

The deployment also establishes operating habits that matter throughout the book: distinguish container startup from application readiness, keep generated credentials private, understand host exposure, preserve named volumes, and verify a useful request rather than trusting process state alone.

Chapter 4, “Run Local Language Models with Ollama,” focuses on the local inference layer. You will manage models, compare practical model sizes, use the generation and chat APIs, and examine CPU and GPU execution in more detail.

Part II: Local Models and Private Chat

Part II follows a prompt through local inference, a stable model gateway, and a browser-based chat interface. You will also learn when a smaller chat stack is a better fit than the complete platform.

Chapter 4: Run Local Language Models with Ollama

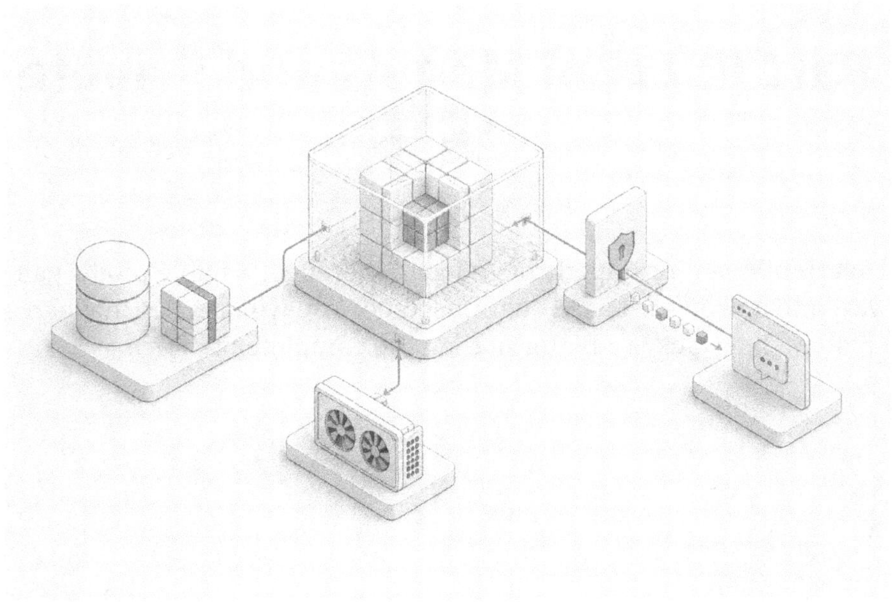

Ollama loads local language models, accepts generation and chat requests, and streams tokens back to a client. This chapter uses InferCrate, the maintained Ollama container in the reference platform, which adds a Bearer-token proxy, persistent model storage, and management commands.

4.1 Deploy InferCrate

If you started the complete platform in Chapter 3 or a maintained lightweight variant, InferCrate is already running and reachable by that stack's GatewayCrate. Its port is not published to the host. Continue with model management below; do not start a second InferCrate container. The standalone steps that follow are for a reader who selected an independent model API in Chapter 2.

For an independent API that you intend to keep on loopback, use the companion repository. For a fresh checkout, run the following commands from its parent directory. If you already have a checkout, enter its current directory and preserve any existing `ollama.env`.

```
git clone https://github.com/hwdsl2/infercrate
cd infercrate
cp ollama.env.example ollama.env
```

The included standalone Compose file publishes port `11434` on all host interfaces. To keep the API host-only, change the mapping before startup:

```
ports:
  - "127.0.0.1:11434:11434/tcp"
```

Render the configuration after any edit:

```
docker compose config --quiet
```

Then start the service:

```
docker compose up -d
```

The container stores all state under `/var/lib/ollama` in the named `ollama-data` volume.

4.2 Retrieve and protect the API key

A fresh persistent deployment generates an API key. Display its endpoint and key for interactive use:

```
docker exec ollama ollama_manage --showkey
```

Retrieve only the key for a script:

```
infer_api_key="$(docker exec ollama ollama_manage --getkey)"
```

Model and management API requests through the maintained image require this Bearer token. The unauthenticated root path is reserved for the container health check. The `ollama_manage` commands run inside the container and do not require the token.

Security: Authentication does not encrypt HTTP. Bind the port to loopback or a trusted interface, and use HTTPS before sending the key across an untrusted network. Do not put a literal key in version control or a shared shell command.

4.3 Download and manage models

List the models already stored in the volume:

```
docker exec ollama ollama_manage --listmodels
```

Pull the small model used by the stack examples:

```
docker exec ollama ollama_manage --pull llama3.2:3b
```

Show models currently loaded in memory and their resource use:

```
docker exec ollama ollama_manage --status
```

Remove a downloaded model by its exact name:

```
docker exec ollama ollama_manage --remove llama3.2:3b
```

Warning: Removal deletes that model from `ollama-data`. It does not delete the whole volume, but the model must be downloaded again before clients can use it.

The optional `OLLAMA_MODELS` environment variable accepts a comma-separated list to pull during initialization. Manual pulls make download timing and failure easier to observe while learning the platform.

The management helper can also re-pull every downloaded model with `ollama_manage --update`. Treat that as a deliberate model update rather than routine container maintenance. A model tag can point to newer weights after the update, so repeat representative quality and resource tests before relying on it in an important workflow.

The environment file also exposes a small set of useful runtime controls:

Setting	What it controls	Operational effect
OLLAMA_MODELS	Models pulled during first initialization	Makes initial downloads automatic but less visible than manual pulls
OLLAMA_MAX_LOADED_MODELS	Models retained in memory at the same time	More loaded models require more RAM or VRAM
OLLAMA_NUM_PARALLEL	Parallel request slots per loaded model	More slots can improve throughput while increasing memory pressure
OLLAMA_CONTEXT_LENGTH	Default context window in tokens	Longer contexts consume more memory and can increase latency
OLLAMA_DEBUG	Verbose diagnostic logging	Useful temporarily; review logs before sharing them

Leave these values unset until a measured workload justifies an override. The current environment example (ollama.env.example) documents the full list and the defaults used by the companion image.

4.4 Choose a model

Choose a model by workload, not by a static catalog. Evaluate:

- Whether it follows the task and produces acceptable output
- Size and quantization of the model weights relative to available RAM or VRAM
- Context length needed by representative prompts
- Latency and throughput at expected concurrency
- Language, coding, tool-use, or structured-output requirements
- Model license and permitted use

The model must leave capacity for the operating system, Ollama's KV cache and other runtime allocations, and the rest of the stack. A model that loads successfully may still become unusable with long contexts or simultaneous requests.

Appendix B provides a reusable model evaluation worksheet. Record the exact model identifier, quantization, context setting, representative inputs, quality failures, latency, resource use, concurrency, and license decision. Treat model recommendations as time-sensitive. Evaluate any candidate not specifically covered in this book against current project documentation and your own workload.

4.4.1 Check whether the first workload is useful

After the selected route and model are running, use one prompt that resembles your actual task, with an expected result you can judge. Time the request with a clock. Use the appropriate request in this chapter for standalone InferCrate or the GatewayCrate request in Chapter 5 for an integrated stack. Record the exact model and context setting, whether the answer meets the task, and the time from request to complete response. For Chapter 2's example, check both the three bullets and the 60-second limit. While it runs, repeatedly sample host available memory with `free -h` and container memory with `docker stats`; on an NVIDIA host, also sample VRAM with `nvidia-smi`. Record the lowest sampled available host memory and the highest sampled container and VRAM use. Samples can miss a peak; Docker's container figures alone do not include all host and GPU allocations.

Compare the observations with the host margin planned in Chapter 2. If the model does not load or memory margin is too small, reduce the service set, context, or model size and retest. If answer quality fails, check the prompt and try a more capable model, then remeasure capacity. If latency fails, test a shorter context or smaller model and check quality again. A successful start is not a capacity result. Add longer-context and concurrent-request measurements before calling this a multi-user deployment. Enter results and the test date in the existing Appendix B worksheets; another host's numbers are not your result.

4.5 Read GGUF and quantization labels

GGUF is a model file format used by llama.cpp-family runtimes. It can hold model tensors and the metadata needed to load them, and Ollama can import supported GGUF artifacts. GGUF is not a model architecture, parameter count, or quality level.

Quantization labels describe how an artifact represents most of its weights. For example, FP16 uses 16-bit floating-point weights, while labels in the `Q8` and `Q4` families identify quantized representations. A lower-bit family generally reduces file size and loaded-weight memory, but the effects on output quality and speed depend on the exact model, quantization method, runtime, and hardware.

Treat the complete label as the artifact's scheme, not as a universal grade. In `Q4_K_M`, for example, `Q4` identifies a 4-bit quantization family and `K_M` identifies a particular llama.cpp scheme and variant. Such artifacts can use more than one tensor representation, so the number is not an exact measure of whole-file bits per weight. Registries and repositories can expose different tag sets, and not every model is available in every scheme.

Model-weight size is still not total runtime memory. The KV cache grows with active context, parallel sequences can add cache capacity, and the runtime needs buffers and implementation overhead. Compare the exact artifact's size and license, then test representative prompts at the intended context and concurrency on the target host. Do not assume that one quantization label is the best default for every model or machine. The current Ollama import documentation and llama.cpp quantization documentation are the authorities when you import or create an artifact.

4.6 Use the generation and chat APIs

The host-side requests in this section apply to the standalone deployment, where port `11434` is published. In the complete and maintained lightweight stacks, InferCrate stays internal; send host application requests through GatewayCrate as shown in Chapter 5.

For the standalone deployment, after setting `infer_api_key`, list the available models:

```
curl -sS http://localhost:11434/api/tags \
  -H "Authorization: Bearer $infer_api_key"
```

Send a nonstreaming generation request:

```
curl -sS http://localhost:11434/api/generate \
  -H "Authorization: Bearer $infer_api_key" \
  -H 'Content-Type: application/json' \
  -d '{
    "model": "llama3.2:3b",
    "prompt": "Explain semantic search in two sentences.",
    "stream": false
```

```
  }'
```

Use the chat API when the request contains roles and conversation turns:

```
curl -sS http://localhost:11434/api/chat \
  -H "Authorization: Bearer $infer_api_key" \
  -H 'Content-Type: application/json' \
  -d '{
    "model": "llama3.2:3b",
    "messages": [
      {"role": "user", "content": "Give one backup rule."}
    ],
    "stream": false
  }'
```

Omit `"stream": false` to receive newline-delimited streaming responses. A client must process multiple response objects and stop when the final object reports completion. The maintained image also exposes an OpenAI-compatible `/v1` subset for compatible clients.

```
unset infer_api_key
```

4.7 Manage persistent model storage

The `ollama-data` volume contains downloaded models, the API key, initialization state, saved port information, and generated proxy configuration. Container removal does not remove this named volume unless a destructive volume option is used.

Model files are often the largest regenerable part of the platform. Monitor storage growth and decide whether backup time and capacity justify copying models or downloading them again after a loss. Preserve the API key and any configuration needed by dependent services.

4.8 Enable GPU acceleration

The standalone CUDA configuration selects `hwdsl2/ollama-server:cuda` and requests an NVIDIA GPU:

```
docker compose -f docker-compose.cuda.yml up -d
```

The current CUDA image supports `linux/amd64` and requires a compatible NVIDIA driver and NVIDIA Container Toolkit. Confirm actual GPU use while a model request is active with `nvidia-smi` and service logs. A healthy container alone does not prove GPU inference.

4.9 Connect InferCrate to other services

Containers in the same Compose network reach InferCrate at `http://ollama:11434`. The complete and lightweight stacks configure GatewayCrate with that internal address and share the generated InferCrate key through a volume.

For a separately deployed GatewayCrate container, add the model with the address that is reachable from that container:

```
docker exec litellm litellm_manage \
  --addmodel ollama/llama3.2:3b \
  --base-url http://ollama:11434
```

Both containers must share a Docker network for the service name `ollama` to resolve. Chapter 5 explains aliases, chat-native routing, and mixed providers.

4.10 Secure and troubleshoot InferCrate

Use this order when a request fails:

1. Confirm the container is running and inspect `docker logs ollama`.
2. Confirm the requested model appears in `ollama_manage --listmodels`.
3. Confirm the key comes from the same persistent deployment.
4. Test the API from the host before testing another container or client.
5. Check memory pressure and `ollama_manage --status`.
6. Confirm the client uses an address reachable from its own network context.

Keep port `11434` unpublished in the complete stack. If direct remote access is required, use a narrow bind address, HTTPS, authentication, and firewall rules.

4.11 Summary and next steps

On a standalone path, InferCrate provides an authenticated local model API with persistent weights and an optional CUDA path. You can manage models independently of containers and test native Ollama and OpenAI-compatible requests. In an integrated stack, host clients use GatewayCrate while InferCrate remains on the Compose network.

Chapter 5 places GatewayCrate in front of this service to provide stable model names, mixed local and hosted routing, virtual keys, and connections to other clients.

Chapter 5: Route Models and Providers with LiteLLM

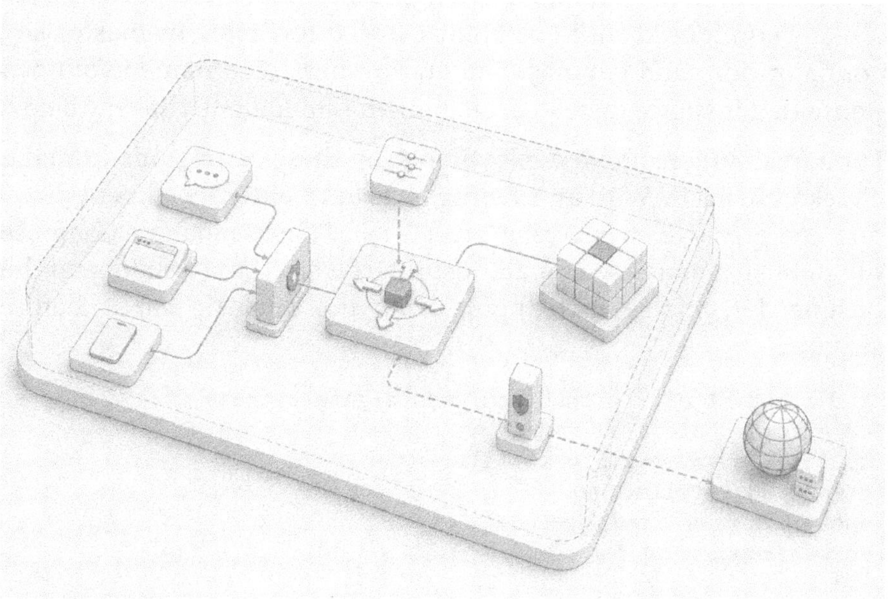

LiteLLM gives applications one OpenAI-compatible model endpoint while routing requests to local model servers or explicitly configured hosted providers. This chapter uses GatewayCrate, the maintained LiteLLM container in the reference platform. It keeps provider URLs, credentials, and model-specific names out of each client.

5.1 Understand the AI gateway role

The gateway sits on a data boundary:

```
AnythingLLM, SDK, coding client, or external runtime
                          |
                          v
                   LiteLLM :4000
                     /           \
             local Ollama    hosted provider
```

LiteLLM normalizes request formats and model names. It does not make every provider equivalent, and it does not keep a request local when the selected route is external. Provider capability, data handling, cost, latency, and failure behavior still apply.

5.2 Deploy GatewayCrate

If you deployed the complete or a lightweight stack, GatewayCrate is already configured with PostgreSQL and InferCrate. Its master key, configuration, and database integration persist in named volumes. Reuse that instance instead of starting the standalone gateway below.

For a fresh independent deployment, run these commands from the checkout parent. With an existing checkout, enter its current directory and preserve any existing `litellm.env`. The standalone Compose file publishes port `4000` on all host interfaces. If only clients on the host need it, change the mapping to `127.0.0.1:4000:4000/tcp` before startup.

```
git clone https://github.com/hwdsl2/gatewaycrate
cd gatewaycrate
cp litellm.env.example litellm.env
chmod 600 litellm.env
docker compose config --quiet
docker compose up -d
```

The one-shot initialization service creates a PostgreSQL password for a fresh database. The database must become healthy before GatewayCrate starts. A proxy with no configured models can be healthy, but it cannot complete a model request.

Retrieve the generated master key:

```
docker exec litellm litellm_manage --showkey
```

5.3 Understand the environment controls

The companion image groups its most consequential environment settings by the connection or capability they enable:

Settings	Purpose
LITELLM_OPENAI_API_KEY	Add an OpenAI credential and the image's current default routes for that provider
LITELLM_ANTHROPIC_API_KEY	Add an Anthropic credential and the image's current default route for that provider
LITELLM_GROQ_API_KEY	Add a Groq credential and the image's current default route for that provider
LITELLM_GEMINI_API_KEY	Add a Google Gemini credential and the image's current default route for that provider
LITELLM_OLLAMA_BASE_URL, LITELLM_OLLAMA_API_KEY	Connect to an Ollama endpoint and supply its API key when required
LITELLM_DATABASE_URL or LITELLM_POSTGRES_PASSWORD_FILE	Enable persistent database features such as virtual keys
LITELLM_MCP_URL, LITELLM_MCP_API_KEY	Register a compatible MCP HTTP endpoint, such as UplinkCrate, on container start
LITELLM_LOG_LEVEL	Select gateway log verbosity

Use the current GatewayCrate environment example (litellm.env.example) for the complete list and current defaults. Store credentials only for connections the deployment is authorized to use.

5.4 Connect InferCrate

The Self-Hosted AI Stack sets `LITELLM_OLLAMA_BASE_URL` to `http://ollama:11434` and shares the InferCrate API key automatically. It maintains two default aliases: `ollama/llama3.2:3b` for compatibility and `ollama-chat/llama3.2:3b` for chat-native behavior such as streaming tool calls.

List effective models:

```
docker exec litellm litellm_manage --listmodels
```

For separate Compose deployments, the service-name URL below works only after both containers join a shared Docker network and GatewayCrate receives the InferCrate key. A host `127.0.0.1` address inside GatewayCrate refers to its own container, not InferCrate. Use `chat-only` for a first combined deployment without this manual wiring. Once the separate network and key are in place, add the model:

```
docker exec litellm litellm_manage \
  --addmodel ollama/llama3.2:3b \
  --base-url http://ollama:11434
```

The command updates persistent `config.yaml` and restarts the proxy. Verify that requests reach InferCrate before using the alias in an application.

5.5 Add hosted model providers

The component environment file supports provider credentials. Store approved keys in `litellm.env`, restrict the file permissions, and recreate the service from the repository root with `docker compose up -d`. When the environment-file path is available, avoid passing a provider key on a shared command line.

Provider model identifiers, retention terms, prices, and the routes added by each credential can change independently of the stack. Use each provider's current official documentation to confirm the account credential and model name, then verify the route with non-sensitive input before authorizing real data. Record the provider, account, region, retention setting, permitted data, budget, and owner in the deployment profile.

Security: A client still calls your self-hosted GatewayCrate endpoint, but a hosted route sends the request content to that provider. Document which aliases are local, which are external, and what data each client may send.

5.6 Configure model aliases and routing

An alias gives clients a stable name while configuration identifies the provider model. Use descriptive names that reveal important boundaries, such as `local-chat` or `approved-hosted-analysis`, rather than disguising an external route as local.

The management helper can add an alias and mark function-calling support. That metadata affects tool-capable clients and must match actual model behavior. Remove a model by the identifier returned by `--listmodels`:

```
MODEL_ID='<model-id>'
docker exec litellm litellm_manage --removemodel "$MODEL_ID"
unset MODEL_ID
```

5.7 Manage virtual keys, budgets, and restrictions

Virtual keys let applications use a scoped key rather than the master key. They require the included PostgreSQL integration.

Create an unrestricted development key:

```
docker exec litellm litellm_manage --createkey
```

Create a key with an alias, selected models, budget, and expiry:

```
docker exec litellm litellm_manage --createkey \
  --alias document-app \
  --models local-chat \
  --budget 20.0 \
  --expires 30d
```

List and revoke virtual keys with `--listkeys` and `--deletekey`. Budgets and model restrictions reduce accidental use, but they do not replace application authorization, network controls, or provider-side limits.

When PostgreSQL integration is active, the LiteLLM Admin UI is available at `http://<server-address>:4000/ui`. Sign in as `admin` with the master key. It is an administrative surface, so keep it on a trusted path and give applications virtual keys instead of sharing that master credential.

5.8 Use the OpenAI-compatible API

The integrated stacks configure the local aliases used below. For a standalone gateway, connect a reachable InferCrate server and verify the alias first; a healthy gateway with no model route cannot answer the example request.

Retrieve the master key into a temporary variable and list models:

```
gateway_master_key="$(docker exec litellm litellm_manage --
  getkey)"

curl -sS http://localhost:4000/v1/models \
  -H "Authorization: Bearer $gateway_master_key"
```

Send a local chat request:

```
curl -sS http://localhost:4000/v1/chat/completions \
  -H "Authorization: Bearer $gateway_master_key" \
  -H 'Content-Type: application/json' \
  -d '{
    "model": "ollama/llama3.2:3b",
    "messages": [
      {"role": "user", "content": "Name one gateway benefit
  ."}
    ],
    "stream": false
  }'

unset gateway_master_key
```

For OpenAI SDK clients, export `OPENAI_BASE_URL` with the value `http://127.0.0.1:4000/v1`. Set `OPENAI_API_KEY` to the GatewayCrate client key. Prefer a scoped virtual key for applications when database integration is active. Preserve the selected model name because it chooses the route. Appendix A lists the component-specific SDK endpoints.

5.9 Connect GatewayCrate to other services

- AnythingLLM uses `http://litellm:4000/v1` inside the stack network and reads the shared GatewayCrate key.
- Coding clients and external runtimes can use the host or HTTPS GatewayCrate URL as an OpenAI-compatible model endpoint.
- UplinkCrate can be registered through `LITELLM_MCP_URL` and a corresponding API key. The stack performs this wiring automatically.
- EmbedCrate has its own OpenAI-compatible endpoint. It is not automatically routed through GatewayCrate in the reference configuration.

List the MCP registrations in GatewayCrate:

```
docker exec litellm litellm_manage --listmcp
```

For a manually managed registration, use `--addmcp <name> <url> --key <key>` and `--removemcp <name>`. These commands update persistent `config.yaml` and restart the proxy, so run them during a controlled change window.

Chapter 16 covers MCP server selection and tool security.

5.10 Secure and troubleshoot GatewayCrate

Protect the master key as an administrative credential. Give applications virtual keys where practical, bind direct HTTP to loopback behind a reverse proxy, and use HTTPS across untrusted networks.

When a request fails, separate the layers:

1. Check `http://localhost:4000/health/liveliness`.
2. Confirm the configured model with `--listmodels`.
3. Identify whether its route is local or hosted.
4. Test the provider independently where appropriate.
5. Check GatewayCrate logs for authentication, database, or provider errors.
6. Confirm a virtual key permits the requested model and remains within limits.

The `litellm-data` volume holds the master key and model configuration. PostgreSQL holds virtual-key and related gateway data. Back up both parts of a stateful deployment.

5.11 Summary and next steps

GatewayCrate provides a stable, authenticated endpoint across selected model routes. Local and hosted providers remain explicit trust boundaries, and virtual keys allow narrower application credentials.

Chapter 6 uses this gateway behind AnythingLLM to create a password-protected browser chat system.

Chapter 6: Build a Private Chat System with AnythingLLM

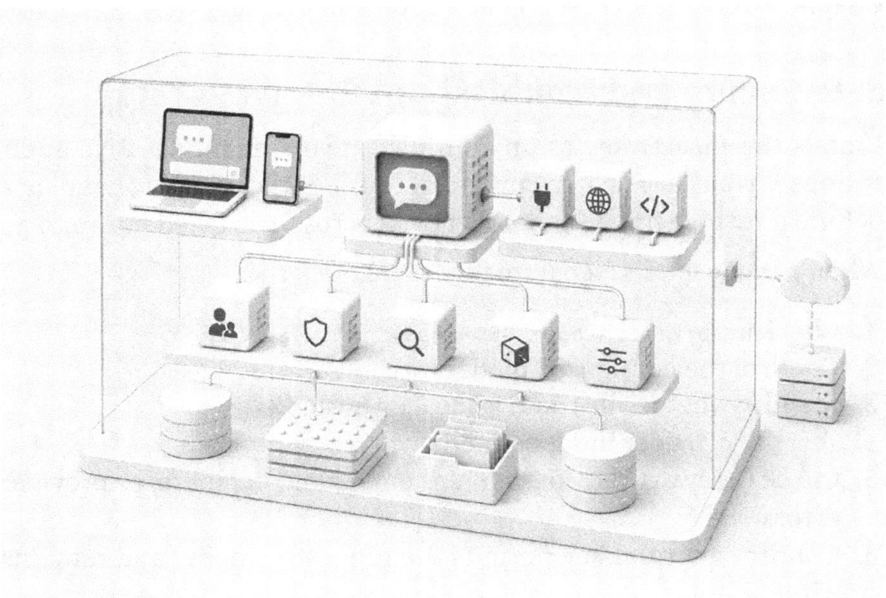

AnythingLLM adds a browser interface, user authentication, conversations, workspaces, and document features to the model gateway. In the maintained chat stack, the bootstrap script connects it to GatewayCrate and creates an administrator password before the first sign-in.

6.1 Understand the chat architecture

```
Browser :3001 --> AnythingLLM --> LiteLLM :4000 --> Ollama
                             |                 |
                             |                 +--> optional hosted
  provider
                             v
                      anythingllm-data
```

AnythingLLM is the interface and application-data layer. LiteLLM selects the model route, and Ollama performs local inference. PostgreSQL serves GatewayCrate and is not AnythingLLM's default vector store. AnythingLLM uses its bundled native embedder and LanceDB unless you deliberately select the stack's EmbedCrate and pgvector options.

Important: The separate EmbedCrate service and pgvector-enabled PostgreSQL are opt-in for AnythingLLM. The default deployment continues to use AnythingLLM's bundled embedder and LanceDB. Back up the workspace before changing either setting, then re-embed previously embedded workspace documents so stored and query vectors use the selected configuration.

6.2 Deploy the chat interface

If you deployed the complete platform in Chapter 3 or `voice-chat` in Chapter 15, AnythingLLM is already running; continue with its connection and credentials below. This section is the book's starting point for `chat-ui`. If you arrived from Chapter 7's variant switch, reuse the existing checkout. From the directory that will contain the checkout, use the following commands; with an existing checkout, begin at `cd self-hosted-ai-stack/stacks/chat-ui`.

```
git clone https://github.com/hwdsl2/self-hosted-ai-stack
cd self-hosted-ai-stack/stacks/chat-ui
docker compose config --quiet
docker compose up -d
docker exec ollama ollama_manage --pull llama3.2:3b
```

From this stack directory, run the repository health-check script (stack-check.sh):

```
../../stack-check.sh
```

The long-running services are InferCrate, GatewayCrate, AnythingLLM, and PostgreSQL. The stack publishes HTTP ports `3001` and `4000` on all host interfaces by default; restrict them to the host or a trusted network before remote access. InferCrate remains internal.

6.3 Connect AnythingLLM to GatewayCrate and InferCrate

The Compose file supplies the GatewayCrate base URL and preferred local model. A shared volume gives the bootstrap script the GatewayCrate key without placing it in the Compose file. On startup, the script prefers the chat-native `ollama-chat/llama3.2:3b` alias when GatewayCrate reports it available and otherwise retains the compatible fallback.

No browser-to-InferCrate connection is required. A chat request reaches InferCrate only after AnythingLLM calls GatewayCrate and GatewayCrate selects the local route.

6.4 Retrieve the initial credentials

Retrieve the seeded administrator password:

```
docker exec anythingllm \
  cat /app/server/storage/.initial_admin_password
```

Open `http://<server-address>:3001`, sign in, and change the password in the security settings.

The initial password file preserves the seeded credential, but it does not change when you update the password in the interface. Treat it as a bootstrap record, not a permanent source of the current password.

GatewayCrate's LiteLLM Admin UI remains available at port `4000`. Use username `admin` and the GatewayCrate master key.

6.5 Work with local and external models

Select a configured GatewayCrate model in AnythingLLM. A local alias routes through GatewayCrate to InferCrate. A hosted alias sends the corresponding prompt and response through the external provider configured in GatewayCrate.

Security: The browser address and self-hosted interface do not determine where model data goes. The selected GatewayCrate route does. Name external routes clearly, restrict application keys where supported, and tell users which content may leave the host.

6.6 Check ordinary and document-based chat

After signing in, use the local `ollama-chat/llama3.2:3b` route and ask a chat question, such as "Give me three ways to organize a workshop." This checks that the browser, gateway, and model can complete a request; it does not test answers from your documents.

For a separate document check, upload and embed a short invented note in the intended workspace using AnythingLLM's document workflow. For example: "Library workshop, Thursday at 6 pm. Registration closes Tuesday. Bring a notebook." Ask, "According to the note, when does registration close? Quote the supporting sentence." Compare the answer with the note itself. AnythingLLM's default workspace retrieval uses its own embedder and LanceDB, not the stack's separate EmbedCrate service or pgvector store. Attaching a file to one chat may use a different path and does not test that workspace retrieval. If the answer is missing or unsupported, confirm that the note is embedded in the intended workspace, then test again. If it remains unsupported, stop using that workspace for document Q&A until the setup is fixed. A plausible answer alone is not evidence that retrieval worked. Chapter 11 explains the services and client logic needed for a separate retrieval pipeline.

6.7 Understand workspaces and data boundaries

AnythingLLM stores its persistent application state under `/app/server/storage` in `anythingllm-data`. This includes its database, workspace and chat state, uploaded or processed material, the persistent `.env` file, and the initial password record.

Workspace separation is an application feature, not a substitute for host, container, or network isolation. Back up the volume before upgrades or model, embedding, or vector-store changes. Switching an embedder or vector store makes previously embedded documents incompatible and normally requires re-embedding.

Do not infer secure deletion from removing an item in the interface. The exact workspace, export, and deletion behavior must be verified against the deployed AnythingLLM version, its vector-store configuration, the persistent volume, and retained backups before making a data-destruction claim.

6.8 Protect the interface

Password protection is enabled for a fresh deployment, but ports `3001` and `4000` use plain HTTP by default. Keep them on the host or a trusted network during setup. For remote or internet-facing access, use the included Caddy overlay or another reviewed reverse proxy that binds direct ports to loopback and adds HTTPS.

A reverse proxy does not replace AnythingLLM authentication or the GatewayCrate master key. Chapter 20 develops the complete exposure checklist.

6.9 Verify and troubleshoot private chat

Verify in dependency order:

1. Confirm InferCrate has the selected model.
2. Confirm GatewayCrate lists the local alias and completes a direct request.
3. Confirm AnythingLLM responds at `/api/ping`.
4. Retrieve the initial password and sign in.
5. Send a short local-model chat message.
6. Recreate the containers and confirm the password and chat state persist.

From the selected lightweight stack directory, useful diagnostics are:

```
docker compose ps --all
docker compose logs --tail=100 litellm
docker compose logs --tail=100 anythingllm
../../stack-check.sh
```

For the complete platform, run `./stack-check.sh` from the repository root instead of the relative command above.

AnythingLLM can start before GatewayCrate is fully ready. During the first deployment, allow initialization to finish before changing configuration in response to a transient error.

6.10 Summary and next steps

When deployed, AnythingLLM provides a persistent, password-protected browser application over the model gateway. Its privacy boundary depends on network exposure, stored workspace data, and the selected GatewayCrate route.

Chapter 7 compares this interface with the smaller API-only chat stack and explains how their shared names and volumes affect migration.

Chapter 7: Choose and Customize a Lightweight Chat Stack

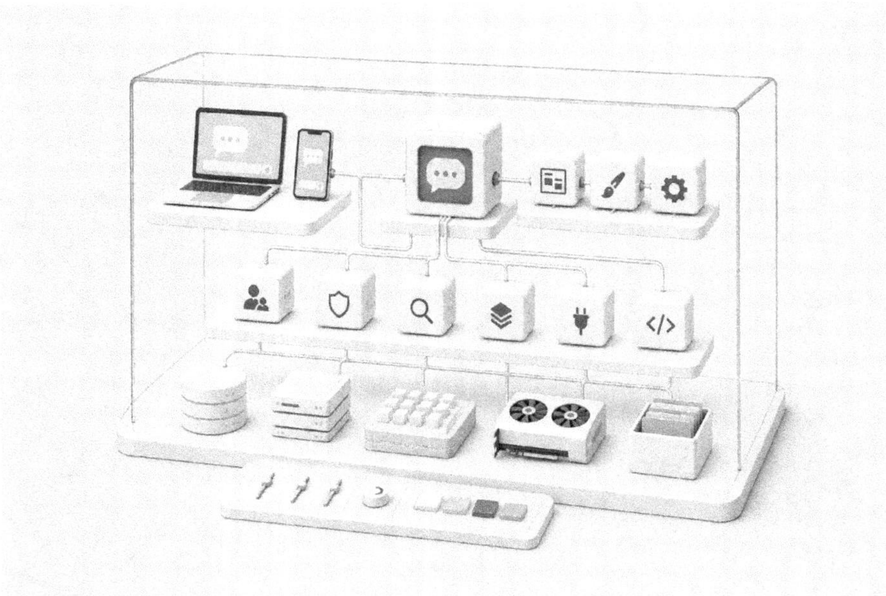

The complete stack is useful for exploration, but many deployments need only a model endpoint or browser chat. The maintained lightweight variants remove services without changing the core InferCrate and GatewayCrate relationship. If you already started `chat-ui` in Chapter 6, use this chapter to compare and customize the variants; do not start another stack. Chapter 6 is the direct starting point for new `chat-ui` readers, while this chapter starts the smaller `chat-only` path.

7.1 Compare chat-only and chat UI configurations

Choice	Services	Starting memory estimate	Interface
chat-only	InferCrate, GatewayCrate, PostgreSQL	About 4.5 GB with a 3B model	API and LiteLLM Admin UI
chat-ui	Above plus AnythingLLM	About 5 GB with a 3B model	Browser chat plus API

These figures are documented estimates, not capacity guarantees. Model size, context, concurrency, and host overhead can dominate the difference.

Choose `chat-only` for SDKs, command-line clients, or another application that already provides its own interface. Choose `chat-ui` when you want password-protected browser chat, workspaces, and document features.

7.2 Deploy the chat-only stack

Run this block only when you chose `chat-only` and no complete or lightweight stack is already running with the default container names.

From the parent directory for the checkout, use the following commands. If you already have the repository, reuse that checkout and begin with `cd self-hosted-ai-stack/stacks/chat-only`.

```
git clone https://github.com/hwdsl2/self-hosted-ai-stack
cd self-hosted-ai-stack/stacks/chat-only
docker compose up -d
docker exec ollama ollama_manage --pull llama3.2:3b
../../stack-check.sh
```

GatewayCrate publishes HTTP port `4000` on all host interfaces by default. Restrict it to the host or a trusted network; InferCrate stays inside the Docker network. The health check verifies PostgreSQL, the model, the gateway key, and routing.

7.3 Switch from chat-only to chat UI

If `chat-ui` already runs, continue below without switching. To move from `chat-only`, stop it without deleting volumes and return to the directory that contains the checkout:

```
docker compose down
cd ../../..
```

Follow Chapter 6's `chat-ui` deployment from there, reusing the checkout and skipping the model pull if it is already present. The named InferCrate, GatewayCrate, and PostgreSQL volumes are shared, so the model and gateway state remain available. The chat UI adds `anythingllm-data` and publishes port `3001`. Retrieve the initial password as described in Chapter 6 before signing in.

7.4 Reduce memory requirements

Start with the smallest service set and model that meet the workload. The most effective reductions usually come from:

- Selecting a smaller or more compressed language model
- Keeping only required models downloaded and loaded
- Avoiding optional AI services until a workflow uses them
- Limiting context length and concurrent requests deliberately
- Choosing `chat-only` when a browser application is unnecessary

Removing a small gateway container while retaining a model that exceeds host memory will not solve the main constraint. Measure the whole workload.

7.5 Add or remove services

The supported stack variants are safer starting points than uncommenting unrelated blocks until a Compose file happens to start. When customizing, trace each service's:

- `depends_on` relationships and readiness requirements
- Network addresses used by other containers
- Published ports and bind addresses
- Persistent and shared volumes
- Generated credentials and env-file mounts

- CPU or CUDA image selection

Render the effective configuration after an edit:

```
docker compose config
```

Chapter 19 develops this method and covers overlays, Podman, and individual containers.

7.6 Move to the complete platform

The lightweight and complete stacks deliberately reuse default container and volume names. A normal transition is:

1. Back up the current named volumes.
2. Run `docker compose down` without `-v` in the current stack directory.
3. Use `cd ../..` to reach the repository root.
4. Render the root configuration with `docker compose config`.
5. Run `docker compose up -d` and then `./stack-check.sh`.

Do not run unmodified variants side by side. Explicit volume names such as `ollama-data` and `litellm-data` are shared globally, while duplicate container names and host ports prevent parallel startup. Deliberate isolation requires a custom naming, port, network, and volume plan.

7.7 Verify and troubleshoot a lightweight stack

Run the repository health-check script (stack-check.sh) from the repository root or use `../../stack-check.sh` from a stack directory. It discovers which services are running and skips services absent from that variant.

If a switch fails, check:

- The previous variant was brought down without deleting volumes.
- Commands are running from the intended stack directory.
- CPU and CUDA Compose contexts are consistent.
- No old container owns a default name or port.
- The shared model and credentials are still present.

7.8 Summary and next steps

The lightweight chat stacks preserve the model and gateway foundation while removing services your workload does not need. Their shared names simplify migration but require one default variant at a time.

Part III adds document conversion, embeddings, reranking, vector storage, and the client-side design needed to assemble a retrieval pipeline.

Part III: Document Intelligence and RAG

Part III builds a retrieval pipeline one observable stage at a time. You will convert documents into structured content, embed and store their chunks, retrieve relevant evidence, and use that evidence to ground model responses. Keeping the stages separate makes quality and authorization failures easier to diagnose.

Chapter 8: Generate Embeddings and Rerank Results

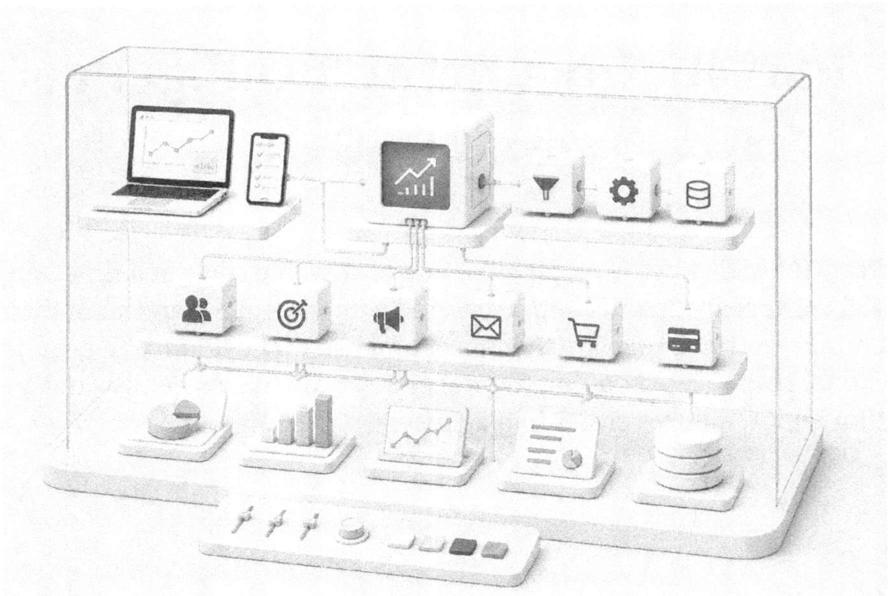

Semantic retrieval lets an application find passages by meaning instead of relying only on exact keywords. In this chapter, you will deploy EmbedCrate, the maintained container powered by Hugging Face Text Embeddings Inference (TEI), convert text into vectors, and optionally rerank candidate passages before sending them to a language model.

8.1 Understand embeddings and semantic retrieval

An embedding model converts text into a fixed-length list of numbers called a vector. Texts with similar meanings tend to produce vectors that are near one another according to a selected distance or similarity measure. An application can therefore embed a question, compare it with stored document vectors, and retrieve passages that are likely to answer the question.

Embedding is only one part of retrieval quality. Parsing, chunking, model choice, metadata filters, and the number of candidates all affect the result. A reranker can improve the final ordering by scoring a smaller set of candidates against the original query with a more expensive model.

The stack separates these responsibilities:

- The embeddings endpoint produces vectors for ingestion and search.
- A vector database stores vectors with their source text and metadata.
- The optional reranking endpoint reorders retrieved candidates.
- The application decides how many results to retrieve, filter, and pass to the language model.

AnythingLLM does not call this separate embeddings endpoint by default. Using it for AnythingLLM workspaces is an opt-in configuration change and requires re-embedding existing workspace documents, as described in Chapter 6.

8.2 Deploy EmbedCrate

The complete platform, both RAG variants, and `code-assistant` include EmbedCrate. If one is running, skip this standalone deployment and use its generated key and published loopback endpoint. Other lightweight variants do not include EmbedCrate; choose a suitable variant or connect this standalone service deliberately rather than starting a duplicate.

The companion repository provides CPU images for common processor architectures. For a fresh standalone deployment, start from the checkout parent:

```
git clone https://github.com/hwdsl2/embedcrate
cd embedcrate
cp embed.env.example embed.env
```

With an existing checkout, enter its current directory and review `embed.env` for model, authentication, offline, and reranking overrides. Standalone Compose publishes port `8000` on all host interfaces; for host-only clients, change it to `127.0.0.1:8000:8000/tcp` before startup.

Start the service and inspect its initial logs:

```
docker compose up -d
docker compose logs --tail=100 embeddings
```

The default model is `BAAI/bge-small-en-v1.5`. Model files and service metadata are stored in the `embeddings-data` volume, so the first start can take longer while the model is prepared.

Within the integrated stack, the service is normally reachable from the host at `http://127.0.0.1:8000` and from other containers by its Compose service name. Keeping the host binding on loopback avoids exposing the endpoint directly to the local network.

A fresh deployment with persistent storage generates an API key automatically. Retrieve it with the bundled management command:

```
docker exec embeddings embed_manage --showkey
```

The most useful configuration settings are those that select models, control offline behavior, or change the service mode:

Setting	Purpose
EMBED_MODEL	Select the Hugging Face embedding model loaded by the server
EMBED_HF_TOKEN	Authenticate downloads of private or gated models
EMBED_LOCAL_ONLY	Disable model downloads and require a populated local cache
EMBED_ENABLED	Disable embeddings when building a rerank-only service
RERANK_ENABLED	Start the optional reranking process
RERANK_MODEL	Select its cross-encoder model
RERANK_PORT	Select the reranking port; it defaults to 8000 in rerank-only mode and 8001 otherwise
RERANK_API_KEY	Set a separate reranker key; when unset, it falls back to EMBED_API_KEY

The current environment example (embed.env.example) is the authority for defaults and the full list. Local-only mode prevents a download attempt; it does not populate the cache for you.

The standalone Compose file mounts `embed.env` by default. An integrated stack does not: create the file beside your selected Compose file and enable its commented `./embed.env:/embed.env:ro` mount before using these overrides.

8.3 Generate embeddings through the API

Set the generated key in a temporary shell variable and submit one or more strings:

```
embed_api_key="$(docker exec embeddings embed_manage --getkey
  )"
curl -sS http://127.0.0.1:8000/v1/embeddings \
  -H "Authorization: Bearer $embed_api_key" \
  -H 'Content-Type: application/json' \
  -d '{
    "model": "BAAI/bge-small-en-v1.5",
    "input": [
      "Private AI keeps selected data on infrastructure you
  control.",
      "Reranking can improve the order of retrieved passages
  ."
    ]
  }'
unset embed_api_key
```

The response follows the OpenAI-compatible embeddings shape. Each item in `data` contains an `embedding` array and an index that corresponds to the submitted input.

The request accepts a `model` field for API compatibility, but the server uses the model selected by its `EMBED_MODEL` configuration. Check the active configuration instead of assuming that the request can switch models:

```
embed_api_key="$(docker exec embeddings embed_manage --getkey
  )"
curl -sS http://127.0.0.1:8000/info \
  -H "Authorization: Bearer $embed_api_key"
docker exec embeddings embed_manage --showinfo
unset embed_api_key
```

8.4 Choose and switch embedding models

Choose an embedding model by considering language coverage, retrieval quality, vector dimension, memory use, model size, and latency. Smaller models are often a practical starting point for a local system. Multilingual or domain-specific collections may justify a different model.

List or download supported models with the management utility:

```
docker exec embeddings embed_manage --listmodels
docker exec embeddings embed_manage --pullmodel BAAI/bge-base
  -en-v1.5
```

After downloading a model, set `EMBED_MODEL` in the mounted `embed.env` and recreate the container from the selected Compose directory. Changing models is a data migration, not merely a runtime setting. Re-embed the existing corpus when the model changes, especially when its vector dimension differs. Query vectors and stored vectors must come from the same compatible embedding space.

8.5 Rerank retrieved documents

Vector similarity is fast enough to search a large collection, but its initial ranking is not always ideal. A reranker reads the query and each candidate together, then produces a relevance score. Use it after vector search on a limited candidate set, not as a replacement for the initial database search.

Set this value in the mounted `embed.env`:

```
RERANK_ENABLED=true
```

In a standalone EmbedCrate deployment, uncomment the reranker port mapping. In an integrated stack, add the mapping to its `embeddings` service if a host client needs the API. With the default reranker port, use `127.0.0.1:8001:8001/tcp` only if host port `8001` is free; the complete platform's optional ScribeCrate Live REST service also uses that host port. If you choose another loopback host port, use it in the request below. A client on the Compose network needs no host mapping. Render the selected Compose file, then recreate the service from that directory:

```
docker compose config --quiet
docker compose up -d --force-recreate embeddings
docker compose logs --tail=100 embeddings
```

The default reranker is `BAAI/bge-reranker-v2-m3`, and its API listens on port `8001`. When `RERANK_API_KEY` is unset, the reranker uses the embeddings key. Submit a query and candidate texts with that default credential wiring. If you configured a separate backend key, use it instead of the embeddings key below.

```
rerank_api_key="$(docker exec embeddings embed_manage --
  getkey)"
curl -sS http://127.0.0.1:8001/rerank \
  -H "Authorization: Bearer $rerank_api_key" \
  -H 'Content-Type: application/json' \
  -d '{
    "query": "How does local retrieval protect data?",
    "texts": [
      "The embeddings endpoint converts text into vectors.",
      "Local retrieval can keep indexed documents inside your
  environment.",
      "A speech model transcribes audio into text."
    ]
  }'
unset rerank_api_key
```

The result includes the original candidate indexes and scores in descending order. Preserve those indexes so your application can reconnect each score to its text and metadata. Set `raw_scores` to `true` only when the application is prepared to interpret unnormalized cross-encoder output. The optional `truncate` field controls whether overlength inputs are shortened to the model limit and defaults to `true` in the documented endpoint.

8.6 Choose a reranking mode

The project supports three useful operating patterns:

- Embeddings only, for the lowest resource use and simplest retrieval path.
- Embeddings plus reranking, for better ordering when quality justifies more latency and memory.

- Reranking only, when another system already provides embeddings and vector search.

Begin without reranking and add it after evaluating real queries. If weak results come from poor parsing or chunking, a reranker will not repair the underlying corpus. GatewayCrate can provide a shared gateway for the reranker. Add a rerank entry to its LiteLLM configuration:

```
model_list:
  - model_name: rerank
    litellm_params:
      model: huggingface/BAAI/bge-reranker-v2-m3
      api_base: http://embeddings:8001
      api_key: os.environ/RERANK_API_KEY
```

The EmbedCrate GatewayCrate integration section documents this route for GatewayCrate's `/rerank` endpoint. Pass the backend's accepted key into the GatewayCrate container as `RERANK_API_KEY`, using its Compose `environment` or a protected `env_file`. The default backend uses the embeddings key; a separately configured reranker uses its own key. Recreate GatewayCrate after changing its environment, then test the route with a GatewayCrate client key. The client key authenticates to the gateway; the backend key authenticates GatewayCrate to the reranker. Direct service calls can simplify initial diagnosis.

8.7 Secure and manage the service

Treat embeddings as derived document data. They can disclose relationships and may still be sensitive even though they do not reproduce the source text directly.

- Keep host ports bound to loopback or behind an authenticated TLS proxy.
- Require an API key and verify upgraded installations explicitly.
- Store secrets outside committed Compose files.
- Restrict access to the model and metadata volume.
- Back up the source corpus and ingestion metadata needed to rebuild the index.
- Do not log complete private documents or embedding requests without a clear need.

The local volume is mainly a model cache and service data store. The vector database introduced in Chapter 10 is responsible for persisted knowledge records.

8.8 Verify and troubleshoot embeddings

Use a short diagnostic sequence before connecting the service to a larger pipeline:

```
docker compose ps
docker compose logs --tail=100 embeddings
embed_api_key="$(docker exec embeddings embed_manage --getkey
  )"
curl -sS http://127.0.0.1:8000/info \
  -H "Authorization: Bearer $embed_api_key"
unset embed_api_key
```

Then send two known texts and confirm that the response contains two vectors of the same nonzero length. Common failures include an incomplete first model download, a wrong API key, a port that was not published, or a client expecting a model different from `EMBED_MODEL`.

If stored vectors fail after a model change, compare the database column dimension with the active model and rebuild the collection. If retrieval quality is weak, inspect source text, chunk size, language coverage, and query phrasing before increasing complexity.

8.9 Summary and next steps

EmbedCrate can create vectors and optionally rerank retrieval candidates. Chapter 9 moves one step earlier in the pipeline by converting PDFs, office files, web documents, and other sources into structured content that is suitable for embedding.

Chapter 9: Convert Documents with Docling

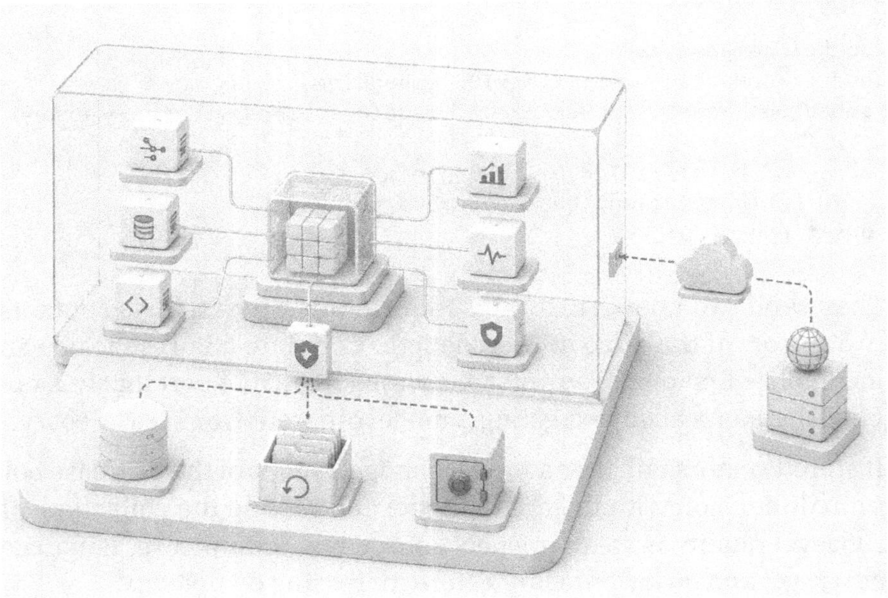

Chapter 8 established how clean text becomes searchable vectors. Real source documents rarely arrive in that form. Useful retrieval begins with useful source text, so this chapter uses Docling and docling-serve through the maintained ParseCrate container to turn uploaded or remotely fetched documents into structured formats suitable for chunking and indexing.

9.1 Prepare documents for AI systems

Copying raw characters from a complex document often loses headings, reading order, lists, and table structure. That damage carries into every later stage. Poorly parsed content produces weak chunks, weak retrieval, and poorly grounded answers.

ParseCrate can process formats such as PDF, DOCX, PPTX, XLSX, HTML, Markdown, LaTeX, AsciiDoc, CSV, and common image formats.

Depending on the request, it can return Markdown, JSON, HTML, plain text, or DocTags. Structured Markdown is a convenient default for inspection and text-oriented ingestion, while JSON is useful when an application needs richer document structure.

9.2 Deploy ParseCrate

`rag-pipeline-full` already runs ParseCrate on a host-loopback port. The complete platform defines it but leaves it disabled; enable that definition and its named volume in your existing Compose file when needed. Other lightweight variants do not include ParseCrate. Follow the standalone block only for an independent service, not to start a second converter beside an integrated one.

For a fresh standalone deployment, start from the checkout parent:

```
git clone https://github.com/hwdsl2/parsecrate
cd parsecrate
cp docling.env.example docling.env
```

With an existing checkout, enter its current directory and review `docling.env` for processing limits, accelerator selection, and authentication. Standalone Compose publishes port `5001` on all host interfaces; for host-only clients, change it to `127.0.0.1:5001:5001/tcp` before startup.

Start the CPU service and inspect its initial logs:

```
docker compose up -d
docker compose logs --tail=100 docling
```

The default API port is `5001`. In the integrated stack it is normally bound to `127.0.0.1:5001`, which keeps it local to the host while remaining available to trusted applications.

ParseCrate includes required models in the image. A persistent volume stores runtime data and optional cache content, but converted documents are normally returned to the client rather than retained as server-side files. The calling application must store any output it needs later.

For a supported NVIDIA host, use the CUDA Compose variant and set `DOCLING_DEVICE` to `cuda` or `auto`. CPU deployment remains the more portable starting point.

Several environment settings materially affect capacity and the accepted input surface:

Setting	Purpose and tradeoff
DOCLING_WORKERS	Adds worker processes for throughput; each worker loads models and increases memory use
DOCLING_ENABLE_UI	Enables the /ui playground; protect it as an additional input surface
DOCLING_MAX_PAGES	Rejects documents above a configured page count
DOCLING_MAX_FILE_SIZE	Rejects uploads above a size in bytes
DOCLING_DEVICE	Selects cpu, cuda, or automatic device detection
DOCLING_LOCAL_ONLY	Disables model downloads and requires any needed cache content to be present locally
DOCLING_LOG_LEVEL	Controls diagnostic verbosity

The current environment example (docling.env.example) contains the full list and current defaults. Application and proxy limits should be no looser than the service limits when documents are untrusted.

9.3 Authenticate requests

ParseCrate uses the `X-Api-Key` request header. Retrieve the generated key with the project management command, then keep it in a temporary shell variable while testing.

```
parse_api_key="$(docker exec docling docling_manage --getkey)
  "
```

Health, readiness, version, and interactive API documentation endpoints do not require the key. Conversion endpoints do. This differs from the Bearer-token pattern used by several other services in the stack, so configure each client deliberately.

9.4 Convert a document from a URL

The source-conversion endpoint accepts a JSON list of remote sources:

```
curl -sS http://127.0.0.1:5001/v1/convert/source \
  -H "X-Api-Key: $parse_api_key" \
  -H 'Content-Type: application/json' \
  -d '{
    "sources": [
      {
        "kind": "http",
        "url": "https://example.com/document.pdf"
      }
    ]
  }'
```

Remote fetching is convenient but creates a server-side request forgery risk. A user who controls the URL may attempt to reach internal services, cloud metadata endpoints, or other destinations available to the ParseCrate container. Production applications should validate schemes and destinations, block private address ranges where appropriate, set timeouts and size limits, and prefer direct uploads when remote fetching is unnecessary.

9.5 Convert an uploaded file

Use the file endpoint when the client already has the document:

```
DOCUMENT_PATH='./document.pdf'
curl -sS http://127.0.0.1:5001/v1/convert/file \
  -H "X-Api-Key: $parse_api_key" \
  -F "files=@$DOCUMENT_PATH"
unset DOCUMENT_PATH
```

The returned object includes converted document content. For a Markdown-oriented pipeline, inspect the response or the deployed API schema to identify the generated Markdown field before passing it to the chunker. This avoids coupling ingestion code to a response path that may change with the Docling Serve version.

9.6 Use synchronous and asynchronous processing

Synchronous conversion keeps the request open until work completes. It is appropriate for smaller documents, command-line testing, and applications whose HTTP timeout exceeds the conversion time.

For larger jobs, submit an asynchronous source request:

```
curl -sS http://127.0.0.1:5001/v1/convert/source/async \
  -H "X-Api-Key: $parse_api_key" \
  -H 'Content-Type: application/json' \
  -d '{
    "sources": [
      {
        "kind": "http",
        "url": "https://example.com/document.pdf"
      }
    ]
  }'
```

The service returns a task identifier. Poll status and fetch the result using that identifier:

```
TASK_ID='<task-id>'
curl -sS "http://127.0.0.1:5001/v1/status/poll/$TASK_ID" \
  -H "X-Api-Key: $parse_api_key"
curl -sS "http://127.0.0.1:5001/v1/result/$TASK_ID" \
  -H "X-Api-Key: $parse_api_key"
unset TASK_ID
```

An application should use bounded polling with backoff, handle failed jobs, and avoid assuming that task results remain available indefinitely.

9.7 Produce structured Markdown

Before indexing a converted document, inspect headings, paragraphs, lists, code blocks, tables, and page boundaries. Remove repeated headers and footers if they pollute the content. Decide how tables should be represented, and retain source identifiers or page references that readers will need for citations.

Do not chunk only by a fixed number of characters if the parsed structure provides better boundaries. Heading-aware chunks are often easier to retrieve and cite. The exact strategy depends on document style. Chapter 11 develops the pipeline design; your ingestion client must implement and test the chosen strategy.

9.8 Understand supported formats and limits

Supported formats and conversion features evolve. Treat the maintained supported-formats documentation and the deployed API schema as the definitive compatibility references for the version you use. Test representative files from your own collection, especially scanned PDFs, complex tables, slide decks, and large files.

Query the running image when you need its exact capabilities:

```
docker exec docling docling_manage --showformats
docker exec docling docling_manage --version
docker exec docling docling_manage --showinfo
```

Relevant controls include maximum file size, maximum page count, worker count, selected processing device, and local-only model behavior. Configure limits before accepting untrusted uploads. Increasing worker count can improve throughput but also multiplies peak CPU and memory demand.

9.9 Manage persistent data

Separate three kinds of data in your design:

- Original source documents, owned by the ingestion application or source repository.
- Converted content, which the client must save if it will be reused.
- Docling runtime data and caches, stored in the service volume.

This distinction matters for backup and deletion. Rebuilding a cache is different from recovering an original document, and deleting a source document does not automatically remove converted text, chunks, embeddings, or database records derived from it.

9.10 Secure the service

Documents are untrusted input, even when the service is private.

- Keep the port on loopback or behind an authenticated TLS reverse proxy.

- Require `X-Api-Key` on conversion calls and rotate exposed credentials.
- Limit file size, page count, concurrency, and request duration.
- Validate remote URLs and restrict outbound access where possible.
- Avoid debug logging that may capture document content.
- Scan or isolate uploads according to the threat model of the environment.
- Define a deletion path for originals and every derived artifact.

If an application sends converted content to a hosted model, the overall system is hybrid even when ParseCrate runs locally. Document that boundary for operators and users.

9.11 Verify and troubleshoot conversion

Check liveness and readiness separately:

```
curl -sS http://127.0.0.1:5001/health
curl -sS http://127.0.0.1:5001/ready
curl -sS http://127.0.0.1:5001/version
docker compose logs --tail=100 docling
```

`/health` reports whether the service process is alive. `/ready` can return HTTP 503 until required models and processing components are ready. An orchestrator should not send conversion traffic based only on the liveness response.

When conversion fails, test a small known-good file, confirm authentication, inspect size and page limits, and review available memory. For poor output, compare another output format and inspect the source document's reading order, embedded fonts, OCR needs, and table complexity.

```
unset parse_api_key
```

9.12 Summary and next steps

ParseCrate provides authenticated local conversion with a clear boundary between original documents, converted content, and runtime caches. Chapter 10 moves downstream from preparation to storage, using pgvector to keep chunks and embeddings together for semantic retrieval.

Chapter 10: Store and Retrieve Knowledge with pgvector

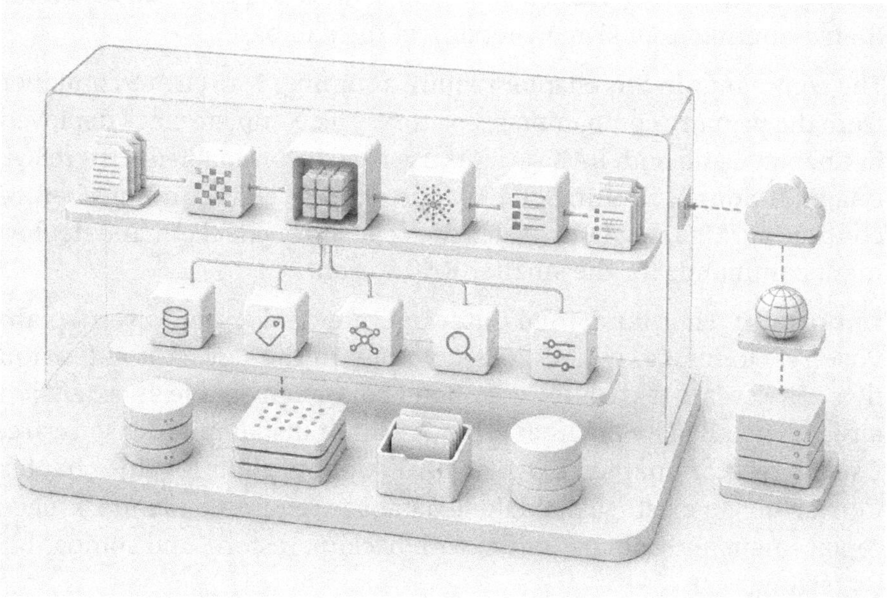

An embeddings API produces vectors, but it does not provide a durable knowledge collection. This chapter enables pgvector in the stack's PostgreSQL service, defines a compact knowledge schema, and outlines safe ingestion and similarity search.

10.1 Understand the role of a vector database

A vector database stores embeddings and finds records whose vectors are nearest to a query vector. For retrieval-augmented generation, each record should also retain the source text and enough metadata to enforce access rules, trace citations, replace changed sources, and delete derived data.

pgvector adds vector types and distance operators to PostgreSQL. It is a practical choice when the deployment already uses PostgreSQL and does not yet need a separate specialized vector service. It also lets an application combine similarity with ordinary SQL filters.

10.2 Understand how pgvector fits into the stack

The RAG variants use a pgvector-enabled PostgreSQL image. PostgreSQL also supports GatewayCrate in the integrated deployment, so knowledge tables should use deliberate names and permissions rather than assuming exclusive ownership of the database.

The commands in this chapter require a running `litellm-db` container from the primary companion repository. The complete stack deployed in Chapter 3 and both RAG variants use a pgvector-enabled PostgreSQL image. If none is running, deploy one of those configurations before continuing. Chapter 11 provides the repository checkout and deployment commands for the smaller RAG variants.

Important: The standalone GatewayCrate deployment uses a plain PostgreSQL image, which does not contain the pgvector extension files. The `CREATE EXTENSION` command below can enable an extension already installed by the image, but it cannot install missing software. Use a pgvector-enabled stack or RAG configuration for this chapter. Converting an existing standalone database requires a planned image replacement and data migration with backup, restore, and compatibility testing.

Enable the extension in the existing `litellm` database:

```
docker exec litellm-db psql -U litellm -d litellm \
  -c 'CREATE EXTENSION IF NOT EXISTS vector;'
```

Confirm that PostgreSQL loaded it:

```
docker exec litellm-db psql -U litellm -d litellm \
  -c "SELECT extname, extversion FROM pg_extension WHERE
  extname='vector';"
```

The pgvector image makes the extension available, but you must still enable it inside each database that uses vector columns.

10.3 Design a minimal knowledge schema

The default `BAAI/bge-small-en-v1.5` model produces 384-dimensional vectors. A minimal table for that model can store stable source identity, chunk order, content, metadata, and the vector:

```
CREATE SCHEMA IF NOT EXISTS knowledge;

CREATE TABLE knowledge.chunks (
    id bigserial PRIMARY KEY,
    source_id text NOT NULL,
    chunk_index integer NOT NULL,
    content text NOT NULL,
    metadata jsonb NOT NULL DEFAULT '{}'::jsonb,
    embedding_model text NOT NULL,
    ingestion_version text NOT NULL,
    embedding vector(384) NOT NULL,
    UNIQUE (source_id, chunk_index)
);
```

The unique source and chunk pair makes reingestion predictable. Store fields such as title, section, page, content hash, tenant, access group, and ingestion version in columns or metadata according to how often they are filtered. Frequently filtered security fields deserve explicit columns and indexes.

10.4 Store text, metadata, and embeddings

An ingestion application should perform these steps as one controlled workflow:

1. Assign a stable identifier to the source.
2. Parse and normalize the document.
3. Split it into ordered chunks.
4. Generate an embedding for each chunk with the selected model.
5. Validate the vector length before writing it.
6. Insert the text, metadata, model identity, and vector.
7. Replace or remove stale records from an earlier source version.

A parameterized insert has this logical shape:

```
INSERT INTO knowledge.chunks (
    source_id,
    chunk_index,
    content,
    metadata,
    embedding_model,
    ingestion_version,
    embedding
)
VALUES ($1, $2, $3, $4::jsonb, $5, $6, $7::vector)
ON CONFLICT (source_id, chunk_index)
DO UPDATE SET
    content = EXCLUDED.content,
    metadata = EXCLUDED.metadata,
    embedding_model = EXCLUDED.embedding_model,
    ingestion_version = EXCLUDED.ingestion_version,
    embedding = EXCLUDED.embedding;
```

Use a database library with bound parameters rather than constructing SQL from document text. Record the embedding model and ingestion version so a later migration can identify records to rebuild.

A production ingestion client must bind all 7 values, reject a vector whose length differs from the schema, and roll back a source update if any chunk fails. A complete ingestion client is outside the scope of this book. The SQL shape above is not a substitute for application-specific transaction and retry tests.

10.5 Run similarity searches

Embed the user's question with the same model, then pass that vector to a parameterized query. Cosine distance uses the `<=>` operator:

```
SELECT
    id,
    source_id,
    chunk_index,
    content,
    metadata,
    embedding <=> $1::vector AS distance
FROM knowledge.chunks
WHERE metadata @> $2::jsonb
ORDER BY embedding <=> $1::vector
LIMIT $3;
```

Smaller cosine distance indicates a closer match. The metadata condition is not merely a relevance feature. In a multi-user system it can be part of the authorization boundary, and it must be derived from trusted identity data rather than arbitrary client input.

Begin with exact search while the collection is small. Approximate indexes can improve performance at larger scale, but index choice and tuning should follow measurements from the real corpus and query load.

Do not choose HNSW or IVFFlat settings without a measured corpus and query load. Record exact-search latency and recall first, then use the selected pgvector project documentation (github.com/pgvector/pgvector) and a repeatable benchmark before adding an approximate index.

10.6 Keep embedding models and dimensions consistent

The vector column dimension, ingestion model, and query model must agree. A 768-element vector cannot be inserted into a `vector(384)` column, and two models with the same dimension still do not necessarily share a meaningful embedding space.

Treat a model change as a versioned reindexing operation:

- Create or select storage for the new dimension and model version.
- Re-embed the authoritative source chunks.
- Direct queries to the matching collection.
- Validate retrieval before retiring the old collection.

Do not silently mix vectors from different models in one searchable collection.

10.7 Define retrieval and data boundaries

For every query, decide which records the caller is allowed to retrieve before ranking them. Useful boundaries include tenant, owner, project, collection, document state, and access group. If a record must never be visible to a caller, filter it in the database query instead of asking the language model to ignore it later.

Deletion must cover all derived records. A stable `source_id` makes it possible to remove chunks when the original is deleted or access is revoked:

```
DELETE FROM knowledge.chunks WHERE source_id = $1;
```

Document retention rules for source files, converted output, chunks, vectors, logs, and backups. Each layer can outlive the one before it unless the system coordinates deletion.

10.8 Protect and maintain the database

Keep PostgreSQL on the internal Compose network unless administrators explicitly need host access. Use generated credentials, a restricted application role, encrypted off-host backups, and tested restoration procedures. Separate knowledge data into its own schema or database when that makes ownership, backup, and permissions clearer.

Because the current stack also uses the database for GatewayCrate, create a least-privilege role for the RAG application instead of reusing the database owner credential. The role needs connection access, schema usage, only the required table operations, and sequence access if the application inserts rows. It does not need permission to administer GatewayCrate tables or create extensions.

Finalize and test the exact grants with the ingestion client before production use. Chapter 22 covers database-consistent backup and restoration; the recovery exercise must confirm both GatewayCrate state and the knowledge schema.

10.9 Summary and next steps

PostgreSQL can now store document chunks and search their embeddings while applying ordinary metadata filters. Chapter 11 joins ParseCrate, chunking, embeddings, pgvector, optional reranking, and GatewayCrate into a complete RAG design.

Chapter 11: Design an End-to-End RAG Pipeline

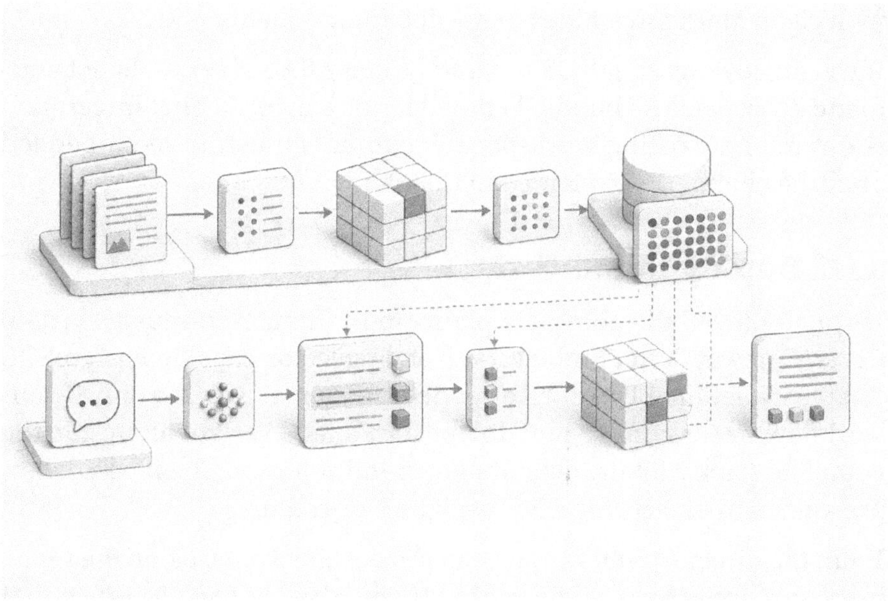

Retrieval-augmented generation, or RAG, retrieves context from your own documents and supplies it to a language model. It can improve relevance and make sources traceable, but it does not guarantee factual answers. This chapter connects the services from earlier chapters into a pipeline whose retrieval and generation stages can be tested separately.

11.1 Choose the basic or full RAG stack

The companion repository provides two starting points:

- The basic RAG stack combines InferCrate, GatewayCrate, the embeddings service, and PostgreSQL with pgvector. It is suitable when source text has already been extracted and normalized.
- The full RAG stack adds ParseCrate for PDFs, office documents, images, and other structured formats.

The documented memory estimate is approximately 5 GB of RAM for the basic stack and 6 GB for the full stack when using the referenced 3B language model. Treat those figures as planning estimates, then measure the selected models, database data, source documents, and real workload on the target host.

Use the basic stack for a text-first prototype. Choose the full stack when conversion is part of the actual product workflow, because testing only with clean Markdown hides many document-quality issues.

If you deploy AnythingLLM separately, these RAG services do not automatically replace its bundled embedder or LanceDB. That integration is opt-in and existing workspace documents must be re-embedded after the change, as explained in Chapter 6.

11.2 Deploy the RAG stack

If you deployed the complete platform in Chapter 3, reuse its InferCrate, GatewayCrate, EmbedCrate, and pgvector-capable PostgreSQL. Enable its optional ParseCrate service and named volume only if your workflow needs conversion; do not start a RAG variant alongside the complete stack with the default names and volumes. The blocks below are alternative starts for a lightweight RAG route.

From the directory where you keep project checkouts, clone the repository and enter its root directory. If you already have it, reuse that checkout and begin with `cd self-hosted-ai-stack`.

```
git clone https://github.com/hwdsl2/self-hosted-ai-stack
cd self-hosted-ai-stack
```

Choose one variant. From the repository root, deploy the basic stack with:

```
cd stacks/rag-pipeline
docker compose up -d
docker exec ollama ollama_manage --pull llama3.2:3b
../../stack-check.sh
```

Alternatively, from the repository root, deploy the full stack with ParseCrate:

```
cd stacks/rag-pipeline-full
docker compose up -d
docker exec ollama ollama_manage --pull llama3.2:3b
../../stack-check.sh
```

These blocks are alternatives. To switch later, run `docker compose down` in the current variant directory, return to the repository root, and then start the other variant. Do not run both with the default names and volumes.

The standard quick start does not require a copied environment file. Component environment files are optional customizations described in the maintained basic RAG (stacks/rag-pipeline/README.md) and full RAG (stacks/rag-pipeline-full/README.md) documentation. Use the same credential and validation practices introduced in Chapter 3, and run only one lightweight stack variant at a time with the default container and volume names.

The deployment commands pull a chat model and run the stack health check. After they complete, confirm the main service boundaries:

- GatewayCrate publishes port `4000` on all host interfaces by default. Restrict access to a trusted network or change the binding before exposure.
- The embeddings API listens on host loopback port `8000`.
- ParseCrate listens on host loopback port `5001` when `rag-pipeline-full` runs or the complete platform's optional service is enabled.
- PostgreSQL and InferCrate normally remain on the internal Compose network.

Enabling the pgvector extension is still required inside the database, as shown in Chapter 10.

11.3 Understand the pipeline

A complete ingestion and question-answering path has distinct stages:

1. Accept and identify a source document.
2. Parse it into structured content.
3. Normalize and split the content into chunks.
4. Embed each chunk and store it with metadata.
5. Embed the user's question.

6. Retrieve allowed chunks by vector similarity and metadata filters.
7. Optionally rerank those candidates.
8. Construct a grounded prompt from the selected context.
9. Generate an answer through GatewayCrate.
10. Return citations and record evaluation signals.

Keeping these stages separate makes failures observable. A fluent but incorrect answer may begin with poor parsing, a missed chunk, an incorrect authorization filter, or a weak prompt rather than with the chat model.

11.4 Parse and chunk source documents

In the full stack, upload a document to ParseCrate and save the returned Markdown or structured JSON in the ingestion application. Do not assume that ParseCrate retains the converted output for you.

Chunk around semantic boundaries such as headings, paragraphs, list groups, or table sections. Each chunk should be large enough to preserve the idea it expresses but small enough to retrieve precisely. Add controlled overlap only when a concept is frequently split across boundaries.

Preserve metadata needed later:

- Stable source identifier and content hash
- Document title and section path
- Page or location reference
- Chunk index
- Tenant or access group
- Parser, chunker, and ingestion versions

No universal chunk size works for every collection. Build a small evaluation set from real questions and adjust the strategy based on missed or noisy passages.

11.5 Generate and store embeddings

Send chunks to `/v1/embeddings` in batches that fit the service and client limits. Confirm that every returned index maps to the correct chunk, validate the vector dimension, and insert the result with a parameterized query like the one in Chapter 10.

An ingestion operation should be repeatable. If a source changes, either replace its chunks in a transaction or write a new version and switch the active collection after validation. Partial updates can otherwise leave old and new content mixed together.

The source repositories demonstrate service calls but do not provide a complete production ingestion application. This chapter defines the required contract but does not include an ingestion program, sample corpus, or application-specific transaction and retry implementation.

11.6 Retrieve relevant context

When a user submits a question, embed it with the same model used for the collection. Execute a similarity search with authorization filters applied in SQL. Retrieve more candidates than you expect to show if you plan to rerank them.

For each result, keep the text, source identity, location metadata, and distance. Inspect retrieved passages before invoking the language model during development. If the required evidence is absent, generation cannot reliably recover it.

The number of candidates returned, often called top-k, is a tradeoff. Too few can omit the answer; too many can add distracting or contradictory context and exceed the model's useful context window. Tune this value against representative queries instead of relying on a universal default.

11.7 Send a grounded prompt through Gateway-Crate

Build a prompt that distinguishes instructions, retrieved context, and the user's question. Assign a short source label to each passage so the model can cite it. A durable instruction pattern is:

```
Answer the question using only the supplied context.
If the context does not support an answer, say that the
  available sources are insufficient.
Cite supporting passages with their source labels.

Context:
[S1] First retrieved passage and its location.
[S2] Second retrieved passage and its location.
```

```
Question:
The user's question goes here.
```

Send that prompt to a configured GatewayCrate model alias rather than tying application code to a specific backend. If the alias uses a hosted provider, the retrieved context crosses the local boundary. Use a local Ollama-backed alias when the content must remain on infrastructure you control.

After generation, verify that every cited label was actually supplied and that the associated passage supports the claim. Model-generated citations are not proof by themselves.

11.8 Add reranking

Reranking fits between vector retrieval and prompt construction:

1. Retrieve a broader set, such as the best 15 or 20 allowed chunks.
2. Send their text and the original question to the reranking service.
3. Reconnect returned indexes to the full records.
4. Keep the best smaller set that fits the prompt budget.

This adds latency and another model to operate. Keep it only when evaluation shows a meaningful gain. Reranking cannot find a passage that the initial retrieval omitted, so its candidate set must be wide enough for the collection.

11.9 Evaluate retrieval and answers

Create a small set of questions with expected source passages and answer criteria. Include answerable questions, questions that require multiple passages, ambiguous questions, and questions that the collection cannot answer.

Evaluate at least two layers:

- Retrieval: Did an allowed supporting passage appear in the candidate set and final context?
- Generation: Did the answer stay within the evidence, represent uncertainty, and cite the right source?

Track changes to parser, chunker, embedding model, retrieval filters, reranker, prompt, and chat model. Without versioning, an apparent quality improvement is difficult to reproduce.

Appendix B provides a compact retrieval evaluation worksheet. Define release gates from the harm and usefulness requirements of the actual workload. For example, an internal discovery tool may tolerate a missed result that a high-impact decision workflow cannot. Never hide per-question authorization or unsupported-answer failures inside one average score.

11.10 Diagnose weak retrieval and unsupported answers

Work from the earliest failing stage:

- If converted text is corrupt, fix parsing or OCR.
- If the answer is split or buried, revise chunk boundaries and metadata.
- If semantically relevant chunks score poorly, evaluate the embedding model and query wording.
- If authorized content is missing, inspect trusted metadata and SQL filters.
- If the right passage is retrieved but not selected, tune top-k or reranking.
- If the right evidence reaches the prompt but the answer exceeds it, strengthen instructions, reduce distracting context, or choose a more suitable chat model.

Log identifiers, versions, distances, and selected source labels for diagnosis. Avoid logging complete private documents or user questions unless policy explicitly permits it.

11.11 Secure the document pipeline

RAG expands the attack and privacy surface because untrusted documents influence a model prompt.

- Treat document text as data, not as trusted instructions. Prompt injection inside a document must not override application policy or tool permissions.

- Apply authorization filters before passages leave the database.
- Restrict remote document fetching and validate uploads.
- Protect API and database credentials separately.
- Make the local or hosted model boundary visible to operators.
- Coordinate deletion across originals, conversions, chunks, vectors, logs, and backups.
- Limit what any later agent or tool can do with retrieved text.

Retrieval improves access to knowledge. It does not grant that knowledge authority to take actions.

11.12 Summary and next steps

The RAG design now covers document conversion, chunking, embedding, storage, retrieval, optional reranking, grounded generation, and evaluation. You still need to implement and verify the application code that connects these stages. Part IV applies the same observable, service-oriented approach to speech recognition and synthesis.

Part IV: Speech and Voice Systems

Part IV adds recorded and real-time speech recognition, local speech synthesis, and the service design behind interactive voice applications. Throughout the pipeline, you will keep audio ownership, client responsibilities, latency, and hosted-model boundaries visible.

Chapter 12: Transcribe Recorded Audio with Whisper

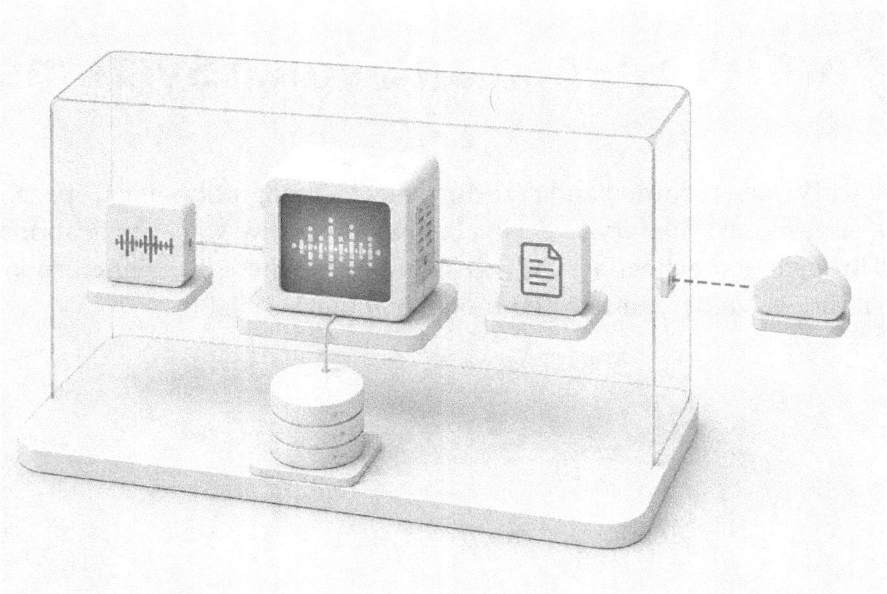

Whisper models turn recorded speech into text. This chapter deploys ScribeCrate, an authenticated, OpenAI-compatible self-hosted transcription and English translation API powered by faster-whisper.

12.1 Choose ScribeCrate or ScribeCrate Live

Use ScribeCrate for complete files such as meetings, interviews, podcasts, and voice notes. Its HTTP API fits jobs where a client can upload a recording and wait for a result. It can also stream decoded segments through server-sent events, but the source is still a submitted file.

Use ScribeCrate Live when audio arrives continuously from a microphone or live stream and partial results must appear with low latency. It uses WebSockets, maintains client sessions, and requires more client logic. Chapter 13 covers that path.

12.2 Deploy ScribeCrate

The complete platform and both voice variants already run ScribeCrate. Reuse their generated key and loopback endpoint instead of following this standalone block. Other lightweight variants do not include it; switch to a maintained voice variant or connect a separate service deliberately.

For a standalone service, start from your checkout parent:

```
git clone https://github.com/hwdsl2/scribecrate
cd scribecrate
cp whisper.env.example whisper.env
```

With an existing checkout, enter its actual directory. Review `whisper.env` for model, language defaults, processing limits, and authentication overrides. The image remains `hwdsl2/whisper-server`; the `whisper` service, environment variables, helper, and volume names are unchanged. Standalone Compose publishes port `9000` on all host interfaces; for host-only clients, change it to `127.0.0.1:9000:9000/tcp` before startup.

Start the default CPU service and inspect its initial logs:

```
docker compose up -d
docker logs --tail=100 whisper
```

The API listens on port `9000`; its default `base` model downloads on first start and is cached in `whisper-data`. Size memory for the model, compute precision and concurrent requests.

On a supported NVIDIA host, use `docker-compose.cuda.yml`. The CUDA image is limited to `linux/amd64` and requires a compatible driver and NVIDIA Container Toolkit. Verify current requirements in the ScribeCrate README before installation.

12.3 Retrieve and protect the API key

Fresh installations with persistent storage generate a Bearer token.

```
docker exec whisper whisper_manage --showkey
```

Use `--getkey` when a script needs only the value. Keep port `9000` on loopback in an integrated stack or place it behind an authenticated HTTPS proxy with upload and rate limits.

12.4 Select and switch transcription models

The default `base` model balances memory, speed, and accuracy. `tiny` reduces resource use, `small` and `medium` can improve accuracy at increasing cost, and the large models target higher quality on capable hardware. English-only variants are optimized for English audio but cannot perform translation.

List and pre-download models before switching:

```
docker exec whisper whisper_manage --listmodels
docker exec whisper whisper_manage --downloadmodel large-v3-
  turbo
```

Set `WHISPER_MODEL` in `whisper.env`. The standalone Compose file mounts that file; an integrated stack requires you to create it beside the selected Compose file and enable its commented `./whisper.env:/whisper.env:ro` mount. From that Compose directory, recreate the service:

```
docker compose up -d --force-recreate whisper
```

Model selection in the request is accepted for compatibility, but the server uses its active configured model.

12.5 Configure transcription behavior

The environment file separates service defaults from per-request choices. The following settings have the greatest operational effect:

Setting	Purpose and tradeoff
WHISPER_LANGUAGE	Uses a known language or automatic detection
WHISPER_DEVICE	Selects CPU, CUDA, or automatic device detection
WHISPER_COMPUTE_TYPE	Selects quantization or precision; the documented defaults are INT8 for CPU and float16 for CUDA
WHISPER_THREADS	Controls CPU inference threads
WHISPER_BEAM	Sets the default decoding beam; larger values can improve accuracy at greater cost

Setting	Purpose and tradeoff
WHISPER_MAX_REQUEST_BEAM	Caps the image's per-request beam extension
WHISPER_MAX_UPLOAD_MB	Rejects oversized uploads with HTTP 413
WHISPER_WORD_TIMESTAMPS	Enables word timing globally; requests can also ask for it individually
WHISPER_LOCAL_ONLY	Disables model downloads and requires a populated model cache

Use the current ScribeCrate environment example (whisper.env.example) for the complete list and defaults. Tune one variable at a time with representative audio because accuracy, latency, and memory interact.

12.6 Transcribe audio

Submit an audio file with the generated key:

```
scribe_api_key="$(docker exec whisper whisper_manage --getkey
  )"
curl -sS http://127.0.0.1:9000/v1/audio/transcriptions \
  -H "Authorization: Bearer $scribe_api_key" \
  -F file=@meeting.m4a \
  -F model=whisper-1 \
  -F language=en
unset scribe_api_key
```

Omit `language` to use automatic detection. A short `prompt` can guide spelling or continuation, but it is not a reliable substitute for post-processing domain terms.

The compatibility API also accepts `temperature`, `response_format`, `stream`, and `timestamp_granularities[]`. The maintained image adds `beam` as a local faster-whisper extension. Check the deployed API schema and documented behavior before using provider-specific options. A client that assumes every hosted transcription option is portable can receive HTTP 400 even though the base OpenAI-compatible route is correct.

12.7 Translate audio

The translation endpoint transcribes supported source languages into English:

```
scribe_api_key="$(docker exec whisper whisper_manage --getkey
  )"
curl -sS http://127.0.0.1:9000/v1/audio/translations \
  -H "Authorization: Bearer $scribe_api_key" \
  -F file=@french-audio.mp3 \
  -F model=whisper-1
unset scribe_api_key
```

Translation requires a multilingual non-turbo model. Output is English, not an arbitrary target language.

12.8 Choose a response format

Use `json` for a simple text property, `text` for plain output, `verbose_json` for timing and segment details, and `srt` or `vtt` for subtitles. Request word timestamps with `verbose_json` only when the downstream application needs them, since detailed output is larger. Add the multipart fields `response_format=verbose_json` and `timestamp_granularities[]=word` to the transcription request when word timing is required.

For a long uploaded file, `stream=true` returns decoded segments through server-sent events. This can improve perceived responsiveness, but it does not create a live microphone session.

12.9 Add speaker diarization

Set `WHISPER_DIARIZATION=true` to enable the optional local sherpa-onnx extension. Its additional models are cached in the same persistent volume and can be pre-downloaded:

```
docker exec whisper whisper_manage --downloaddiarize
```

With `verbose_json`, segments gain speaker labels such as `SPEAKER_00`. Diarization runs after transcription and is skipped for streaming responses. It is a local extension, not an implementation of provider-specific diarization models or fields. The maintained speaker diarization instructions contain the current model, environment, and response details.

12.10 Manage CPU and GPU performance

Latency depends on audio duration, model size, compute device, quantization, beam size, CPU threads, and concurrent requests. Begin with the default model and one request, then measure representative audio. On CPU, appropriate thread count and INT8 compute are practical defaults. On CUDA, the project uses a GPU-oriented compute type.

Do not estimate capacity from model loading alone. Test overlapping uploads, time to first streamed segment, total processing time, and memory peaks.

12.11 Alternative native Linux deployment

If a supported Linux host should run Whisper directly under systemd instead of Docker, the Whisper installation project (github.com/hwdsl2/whisper-install) provides an automated native deployment with the same general API style. Its service account, configuration paths, logs, updates, and rollback process differ from the container workflow. The rest of this book uses Docker Compose so shared networks, volumes, and credentials remain consistent across the platform.

12.12 Secure and troubleshoot ScribeCrate

- Treat audio and transcripts as sensitive user data.
- Restrict upload size and accepted workflows at both proxy and application layers.
- Keep debug logging off when it could capture transcript text.
- Rate-limit expensive endpoints and limit concurrency.
- Preserve the model volume if download time matters, but back up original audio and retained transcripts separately.

Check `docker logs whisper`, `docker exec whisper whisper_manage --showinfo`, and `/v1/models` when a client fails. A first request may wait for model loading. Format errors often come from the media container or codec, while memory failures usually require a smaller model or more resources.

12.13 Summary and next steps

ScribeCrate provides authenticated file transcription, English translation, subtitles, timestamps, and optional speaker labels. Chapter 13 changes the delivery pattern from uploaded files to live audio sessions with ScribeCrate Live.

Chapter 13: Stream Real-Time Transcription with WhisperLive

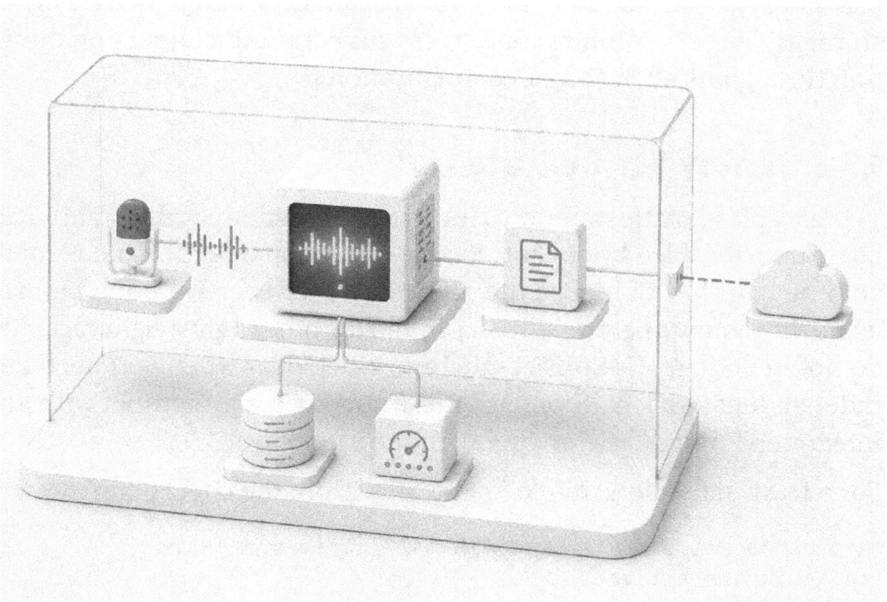

ScribeCrate Live uses WhisperLive and faster-whisper for authenticated WebSocket transcription and speech-to-English translation. It supports live captions and microphone-driven interfaces, plus REST file transcription.

13.1 Choose ScribeCrate Live instead of batch transcription

Choose ScribeCrate Live when a workload needs transcription while audio is still arriving. Choose ScribeCrate when the input is a complete recording and simpler HTTP processing is more important than interactive latency.

Real-time transcription adds connection state, audio conversion, silence handling, reconnection, and concurrency limits. If file transcription meets the requirement, the batch service is easier to run.

13.2 Understand the real-time architecture

A client sends mono float32 PCM at 16 kHz, normalized to `[-1, 1]`, as binary WebSocket frames. A JSON handshake selects the task and source language; the server returns timed text segments and completion state.

The client owns microphone permission, resampling, buffering, display of partial results, recovery from disconnects, and final transcript storage. The server limits concurrent sessions and connection duration through its environment configuration.

13.3 Deploy ScribeCrate Live

The complete platform and `voice-pipeline` define ScribeCrate Live but leave it disabled by default. Enable the optional service and named volume in either Compose file, then verify its effective ports before using the requests below. `voice-chat` and the other lightweight variants do not include it. Use this standalone block only when you need an independent service; its default REST port `8000` may conflict with an integrated EmbedCrate endpoint on the same host.

For a fresh standalone deployment, start from the checkout parent:

```
git clone https://github.com/hwdsl2/scribecrate-live
cd scribecrate-live
cp whisper-live.env.example whisper-live.env
```

With an existing checkout, enter its current directory and review `whisper-live.env` for models, connection limits, session duration, and authentication overrides. Standalone Compose publishes ports `9090` and `8000` on all host interfaces; for host-only clients, bind both to `127.0.0.1` before startup.

Start the CPU service and inspect its initial logs:

```
docker compose up -d
docker logs --tail=100 whisper-live
```

Standalone port `9090` serves WebSockets and host port `8000` serves REST. In the complete platform and `voice-pipeline`, the optional service binds WebSocket host port `9090` and REST host port `8001` to loopback. Other containers should use its service name and container ports instead of host mappings.

Use the CUDA Compose file on a supported NVIDIA `linux/amd64` host. Models are cached in the `whisper-live-data` volume and normally load on the first client connection.

The settings most likely to affect a real-time deployment are:

Setting	Purpose and tradeoff
WHISPERLIVE_MODEL	Sets the WebSocket model and REST default
WHISPERLIVE_MAX_CLIENTS	Caps simultaneous WebSocket sessions
WHISPERLIVE_MAX_CONNECTION_TIME	Disconnects sessions after a configured number of seconds
WHISPERLIVE_THREADS	Controls CPU inference threads and non-GPU work
WHISPERLIVE_LOCAL_ONLY	Disables model downloads and requires a populated cache

Use the current ScribeCrate Live environment example (whisper-live.env.example) for port, log, authentication, and offline settings as well as the complete defaults. Pre-download a transcription model before a planned session when first-use latency is unacceptable:

```
docker exec whisper-live whisper_live_manage --listmodels
docker exec whisper-live whisper_live_manage \
  --downloadmodel large-v3-turbo
```

After changing `WHISPERLIVE_MODEL` in `whisper-live.env`, recreate the service. Standalone Compose mounts that file; in the complete or `voice-pipeline` stack, create it beside the selected Compose file and enable its commented `./whisper-live.env:/whisper-live.env:ro` mount. From that Compose directory, recreate the service and confirm the active selection:

```
docker compose up -d --force-recreate whisper-live
docker exec whisper-live whisper_live_manage --showinfo
```

13.4 Authenticate the WebSocket connection

Fresh persistent installations with the data volume generate an API key automatically. Display it for an interactive client setup with:

```
docker exec whisper-live whisper_live_manage --showkey
```

Use `whisper_live_manage --getkey` when a script must capture only the key. Keep the value in memory or a secret store for as little time as practical, and unset temporary shell variables after use.

Authentication happens during the WebSocket handshake, before the client sends its configuration message. A non-browser client can send `Authorization: Bearer <your-api-key>` as a handshake header or use the token query parameter. The browser WebSocket API cannot set an arbitrary `Authorization` header, so browser clients use a URL such as:

```
wss://ai.example.com/ws?token=<your-api-key>
```

Use `wss://` whenever the connection crosses an untrusted network. Treat a token in a query string as a secret because proxies and diagnostics can record request URLs. Redact it from access logs and shorten its lifetime when an application can issue session-specific credentials.

13.5 Understand the WebSocket protocol

ScribeCrate Live uses WhisperLive's WebSocket protocol. After connecting, send configuration before audio:

```
{
  "uid": "browser-client-1",
  "language": "en",
  "model": "base",
  "task": "transcribe",
  "audio_format": "float32",
  "use_vad": true
}
```

WebSocket uses server-side `WHISPERLIVE_MODEL`; the client's `model` field does not override it. Use `"use_vad": true` for voice activity detection, or `false` to disable it.

Wait for `SERVER_READY`, then send raw PCM without WAV headers or compression. A result resembles:

```
{
  "uid": "browser-client-1",
  "segments": [
    {
      "text": "Hello, how are you?",
      "start": 0.0,
      "end": 2.4,
      "completed": true
    }
  ]
}
```


Clients should distinguish provisional from completed text and deduplicate segments according to the protocol behavior observed in the deployed version.

For speech-to-English translation, set `"task": "translate"` and select the source language, for example `"language": "es"`; use `"language": null` for detection. Output is English, not an arbitrary target language. Choose a multilingual non-turbo server model; English-only `.en` models cannot translate, and turbo is not trained for this task.

13.6 Transcribe microphone and file input

Use the maintained authenticated Python client example, which adds a Bearer header during the WebSocket handshake and retains file and microphone input. Install the client version specified by that example in an isolated application environment, then export the server key:

```
export WHISPERLIVE_API_KEY="$(docker exec whisper-live
  whisper_live_manage --getkey)"
```

The authenticated Python example accepts a file path for recordings or no arguments for microphone capture. If authentication is deliberately disabled, set `WHISPERLIVE_API_KEY` to an empty string. Keep the client separate from the server container; test its audio device, sample rate, reconnection, and shutdown. Unset the exported key after use.

13.7 Connect a browser client

A browser can create a WebSocket, send the configuration object, capture audio with `getUserMedia`, resample it to the required PCM format, and send binary chunks. Use `wss://` for any nonlocal page. Browsers restrict microphone access to secure contexts, and the proxy should validate WebSocket origins before upgrading the connection.

Pass the key into application code at runtime rather than committing it in the page source:

```
const apiKey = "<your-api-key>";
const url =
  `wss://ai.example.com/ws?token=${encodeURIComponent(apiKey)
  }`;
const ws = new WebSocket(url);
```

Do not place a long-lived shared API key in public browser code. Prefer a backend that authenticates the user and issues a short-lived session or proxies the stream under application authorization. A complete browser capture and audio-conversion client is outside the scope of this book.

13.8 Use the REST API

The REST service accepts OpenAI-compatible file transcription requests at `/v1/audio/transcriptions`. It is useful for compatibility testing or when one deployment needs both live and uploaded inputs. For a file-only workload, the smaller ScribeCrate service remains the clearer choice.

The request below uses the standalone host mapping `8000`. When enabled in the complete platform or `voice-pipeline`, the default REST host mapping is `127.0.0.1:8001:8000`; use host port `8001` there instead. A client on the Compose network uses `whisper-live:8000` rather than either host port.

ScribeCrate Live serves `/docs`, `/redoc`, `/openapi.json`, and `/health` without authentication. REST transcription still requires the Bearer header when authentication is enabled, including requests from Swagger UI. The WebSocket `token` query parameter does not authenticate REST requests.

```
scribe_live_api_key="$(docker exec whisper-live \
  whisper_live_manage --getkey)"
curl -sS http://127.0.0.1:8000/v1/audio/transcriptions \
  -H "Authorization: Bearer $scribe_live_api_key" \
  -F file=@meeting.m4a \
```

```
  -F model=whisper-1 \
  -F language=en
unset scribe_live_api_key
```

Omit `model` or use `model=whisper-1` for `WHISPERLIVE_MODEL`; a supported model size selects a different model for that request.

13.9 Tune latency and resource use

Smaller models, GPU inference, shorter client buffers, and voice activity detection can reduce latency. Extremely small buffers create overhead and unstable partial results, while large buffers delay visible text. Measure the full path from microphone capture to rendered transcript.

Set explicit maximum clients and connection duration. Test simultaneous sessions because one client can appear responsive even when the host cannot sustain the intended concurrency.

13.10 Secure and troubleshoot streaming

The generated key protects REST transcription and WebSocket connections. REST uses a Bearer token; WebSocket clients can use the supported header or query parameter. When authentication is enabled, verify after upgrades that both interfaces reject missing or invalid credentials and accept valid ones.

Bind the ports to loopback behind a reverse proxy even when native authentication is active. Never expose an unauthenticated compute endpoint directly.

For connection failures, verify the `Upgrade` and `Connection` proxy headers, TLS scheme, origin policy, token handling, PCM format, and server client limits. For CUDA failures, confirm the container can see the GPU before tuning the model.

13.11 Summary and next steps

ScribeCrate Live supplies the live input side of a voice application. Chapter 14 completes the other end of the interaction by generating spoken output through a local SpeakCrate text-to-speech API.

Chapter 14: Generate Speech with Kokoro

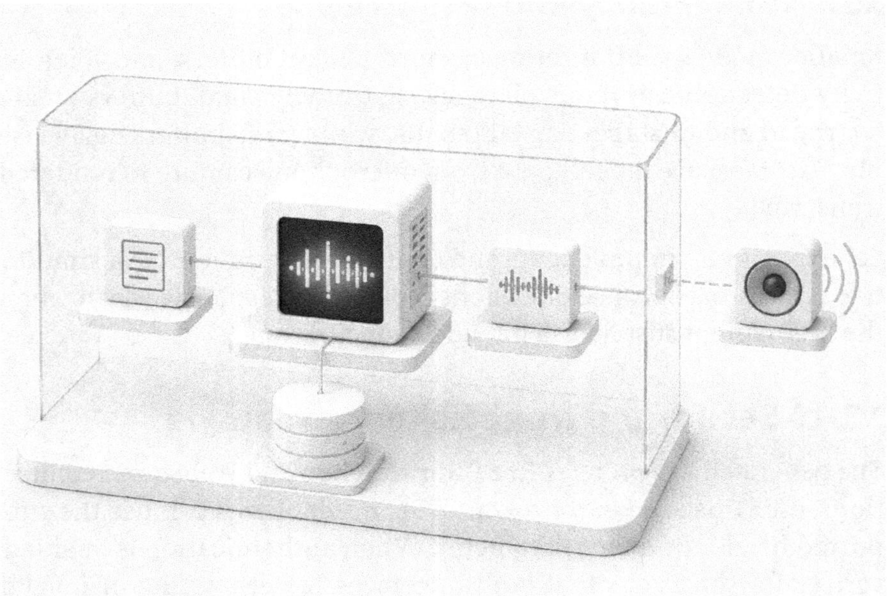

Kokoro generates speech from text using local voices and model files. This chapter deploys SpeakCrate, the maintained Kokoro container that provides an OpenAI-compatible speech API for applications and voice pipelines.

14.1 Deploy SpeakCrate

Both voice variants already run SpeakCrate. The complete platform defines it but leaves it disabled until you enable the optional service and named volume. Other lightweight variants do not include it. Reuse or enable the service in your selected Compose project before considering this independent deployment block.

For a fresh standalone deployment, start from the checkout parent:

```
git clone https://github.com/hwdsl2/speakcrate
cd speakcrate
cp kokoro.env.example kokoro.env
```

With an existing checkout, enter its current directory and review `kokoro.env` for voice, speech speed, language pipeline, offline, and authentication overrides. Standalone Compose publishes port `8880` on all host interfaces; for host-only clients, change it to `127.0.0.1:8880:8880/tcp` before startup.

Start the CPU service and inspect its initial logs:

```
docker compose up -d
docker logs --tail=100 kokoro
```

The API listens on port `8880`. The shared model is downloaded and cached in `kokoro-data` on first start. Use the CUDA Compose file on a supported NVIDIA `linux/amd64` host.

Set shared defaults in the environment file instead of each client:

Setting	Purpose
KOKORO_VOICE	Selects the default voice; an individual request can override it
KOKORO_SPEED	Selects the default speaking speed
KOKORO_LANG_CODE	Preloads one language pipeline; other pipelines can still load on demand
KOKORO_LOCAL_ONLY	Disables model downloads and requires a populated cache

The current SpeakCrate environment example (kokoro.env.example) contains the complete list and current voice-to-language behavior.

14.2 Retrieve and protect the API key

Fresh persistent installations generate a Bearer token.

```
docker exec kokoro kokoro_manage --showkey
```

Keep the endpoint on loopback or behind an authenticated HTTPS proxy. Rate limits matter because a public text-to-speech endpoint can consume sustained CPU or GPU time.

14.3 Select a voice

List voices from the running version instead of copying a catalog that may become stale:

```
docker exec kokoro kokoro_manage --listvoices
```

Voice identifiers encode language and voice. The service selects the corresponding language pipeline from the identifier. OpenAI voice names are accepted as compatibility aliases, but they map to local Kokoro voices and do not reproduce proprietary voices.

Evaluate voices with the language, names, punctuation, numbers, and sentence lengths that the application will actually speak. A pleasing demo sentence is not enough to choose a production voice.

14.4 Synthesize speech through the API

Keep input within the deployed service's per-request limit. Save the binary response instead of treating it as JSON:

```
speak_api_key="$(docker exec kokoro kokoro_manage --getkey)"
curl -sS http://127.0.0.1:8880/v1/audio/speech \
  -H "Authorization: Bearer $speak_api_key" \
  -H 'Content-Type: application/json' \
  -d '{
    "model": "tts-1",
    "input": "Your private voice service is ready.",
    "voice": "af_heart",
    "response_format": "mp3",
    "speed": 1.0
  }' \
  --output speech.mp3
unset speak_api_key
```

Supported output options include MP3, Opus, AAC, FLAC, WAV, and raw PCM. The API also supports chunked audio or server-sent-event streaming.

The API reference defines accepted values and limits for these controls:

Field	Behavior
speed	Selects the requested speaking speed

Field	Behavior
response_format	Selects mp3, opus, aac, flac, wav, or headerless signed 16-bit PCM at 24 kHz mono
stream_format	Selects chunked audio or server-sent events with audio delta and completion events
volume_multiplier	Scales output volume with clipping protection
instructions	May be accepted without affecting synthesis; verify engine support.

For input longer than the limit, split at sentence-aware boundaries. Decide whether to retry or skip a failed segment, or stop the complete synthesis job.

14.5 List voices and models

Use the live discovery endpoints when configuring a client:

```
speak_api_key="$(docker exec kokoro kokoro_manage --getkey)"
curl -sS http://127.0.0.1:8880/v1/voices \
  -H "Authorization: Bearer $speak_api_key"
curl -sS http://127.0.0.1:8880/v1/models \
  -H "Authorization: Bearer $speak_api_key"
docker exec kokoro kokoro_manage --showinfo
unset speak_api_key
```

The model names accepted by the compatibility API refer to the same local Kokoro model. Do not infer provider-specific quality tiers from names such as `tts-1-hd`.

14.6 Change the voice and defaults

All included voices use the same shared model file. Set `KOKORO_VOICE` in `kokoro.env`. Standalone Compose mounts it; in a complete or voice stack, create it beside the selected Compose file and enable its commented `./kokoro.env:/kokoro.env:ro` mount. From that Compose directory, recreate the service. Individual requests can still specify another voice.

```
docker compose up -d --force-recreate kokoro
```

Changing voices does not require another model download. Recheck the maintained SpeakCrate project documentation before claiming support for another model family.

14.7 Manage generated audio and persistent data

The service volume stores model cache and runtime metadata. The client owns generated audio. Save, stream, expire, or delete it according to the application policy. Backing up `kokoro-data` avoids a model re-download but does not recover output files stored elsewhere.

Input text may be as sensitive as a transcript or document excerpt. Keep debug logs from recording it on shared systems.

14.8 Tune performance

Measure time to first audio, total synthesis time, real-time factor, memory, and behavior under concurrency. CPU is a simple default; CUDA may improve throughput on a compatible host.

14.9 Secure and troubleshoot SpeakCrate

Limit request body size, maximum text length, concurrency, and rate. Validate client-selected voices and formats instead of passing arbitrary values. For browser use, configure TLS, authentication, and CORS at the application or proxy layer.

If output is empty or invalid, inspect service logs, confirm the key, try a known voice from `/v1/voices`, and save the response without piping it through a text processor. If pronunciation is poor, test punctuation, language-specific voices, and shorter sentence units.

14.10 Summary and next steps

Chapter 15 combines transcription, model inference, and speech output in a browser-based stack with separate voice APIs or a headless pipeline.

Chapter 15: Assemble Voice AI Stacks and Design Voice Pipelines

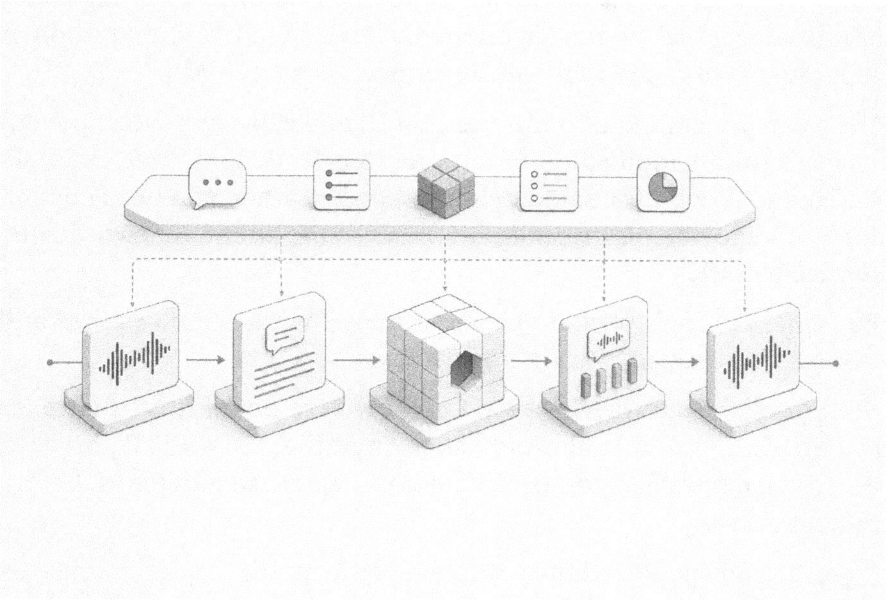

A complete voice interaction transforms audio into text, text into a model response, and that response into new audio. One stack variant places a chat interface alongside the speech services, while another provides headless APIs for custom applications. Colocating these services does not automatically connect voice controls in the browser interface.

15.1 Understand the voice pipeline

The core path is:

```
Audio input -> Whisper -> transcript -> LiteLLM -> Ollama ->
  response -> Kokoro -> audio output
```

LiteLLM can route the response stage to a hosted model instead, but doing so sends the transcript and prompt beyond the local deployment. The application owns recording, session state, playback, and the decision about which artifacts to retain.

15.2 Choose voice chat or a headless voice pipeline

The voice chat stack adds AnythingLLM for a web interface, workspaces, and saved conversations. It is the closer starting point for interactive use, with a documented estimate of about 6.5 GB of RAM using the referenced 3B model. A verified UI configuration or separate client is still required to connect its speech APIs.

The voice pipeline stack omits the chat UI and exposes the component services for an application or automation. Its documented estimate is about 6 GB with the same model. Choose it when you need custom session logic, telephony integration, a mobile client, or a service-to-service workflow.

Run only one lightweight variant at a time with default names and volumes because the variants share resources.

Both publish GatewayCrate's HTTP port `4000` on all host interfaces by default; `voice-chat` also publishes AnythingLLM's HTTP port `3001`. Restrict each published port to the host or a trusted network before remote access.

If you already deployed the complete platform in Chapter 3, reuse its ScribeCrate, GatewayCrate, InferCrate, and AnythingLLM services. Enable its optional SpeakCrate service and named volume for speech output. The variant startup blocks below are alternatives for a smaller voice-specific deployment, not additions to the running complete platform. `voice-chat` includes a browser UI; `voice-pipeline` does not.

Both variants are maintained in the companion repository. From the directory where you keep project checkouts, clone it before choosing a variant:

```
git clone https://github.com/hwdsl2/self-hosted-ai-stack
```

If you already have the repository, reuse that checkout. Each alternative deployment block below assumes your shell starts in the directory that contains the `self-hosted-ai-stack` checkout.

15.3 Deploy the voice chat stack

```
cd self-hosted-ai-stack/stacks/voice-chat
docker compose up -d
docker exec ollama ollama_manage --pull llama3.2:3b
../../stack-check.sh
```

Open AnythingLLM on port `3001` and use the generated password described in Chapter 6. The stack wires its GatewayCrate model connection automatically. It does not configure AnythingLLM to call ScribeCrate and SpeakCrate automatically. Those services remain on host loopback by default and are available for a supported UI configuration or a separate client that implements the three-stage flow.

15.4 Deploy the headless voice pipeline

If you started the voice chat variant, stop it from that stack directory before switching:

```
docker compose down
cd ../../..
```

The second command returns to the directory that contains the repository checkout. Run the headless alternative from there:

```
cd self-hosted-ai-stack/stacks/voice-pipeline
docker compose up -d
docker exec ollama ollama_manage --pull llama3.2:3b
../../stack-check.sh
```

Use the CUDA Compose file when the host meets the GPU requirements. ScribeCrate Live is present as an optional commented service and must be enabled deliberately for live streaming.

15.5 Connect ScribeCrate, GatewayCrate, and SpeakCrate

The maintained voice pipeline example (stacks/voice-pipeline/README.md) includes a shell demonstration of all three APIs. For a production client, keep JSON construction out of string interpolation so transcript content cannot break the request. The logical flow is:

1. Retrieve the three generated API keys.

2. Upload audio to ScribeCrate and read the returned `text` value.
3. Build the request with a JSON serializer and send it through GatewayCrate.
4. Extract the assistant response.
5. Serialize a SpeakCrate request and save or stream the returned audio.

The linked README example is useful for a trusted sample, but direct shell interpolation is not a safe general-purpose implementation for arbitrary transcript or model text. A production three-stage client is outside the scope of this book. Any implementation must add safe JSON serialization, bounded timeouts, stage-level retries, and sanitized error handling.

15.6 Add real-time transcription

Enable ScribeCrate Live when the client must display partial text or begin processing before the speaker finishes. The client still must buffer completed segments and decide when an utterance is ready for the language model. Sending every partial update creates duplicate requests and unstable responses.

Run `whisper_live_manage --getkey` after enabling the service to retrieve its generated key. A non-browser WebSocket client can authenticate with a Bearer header or the `token` query parameter. Browser clients can use the query parameter because the browser WebSocket API cannot add an arbitrary `Authorization` header. Use WSS, avoid long-lived keys in browser code, and redact token-bearing URLs from proxy logs.

A conversational client also needs an interruption policy. Decide whether new speech cancels current generation or playback, queues another turn, or is ignored until the response finishes.

15.7 Test each stage independently

Before testing the whole interaction:

1. Transcribe a known audio file and inspect the text.
2. Send a fixed text prompt through GatewayCrate and inspect the model response.
3. Synthesize a fixed sentence with SpeakCrate and play the file.

4. Pass the known transcript into the model stage.
5. Pass a known model response into the speech stage.
6. Run the full path and record latency for every boundary.

This sequence shows whether a failure belongs to media decoding, recognition, model routing, generation, synthesis, or the client.

15.8 Secure and troubleshoot the pipeline

Protect every service key, not only the public entry point. Keep component ports internal or on loopback, and expose only the application or authenticated proxy. Apply upload limits, rate limits, session duration, and concurrency controls.

Voice data can reveal identity, health, location, and private conversation. Define retention for raw audio, transcripts, prompts, model responses, generated speech, and logs. Make hosted model routing visible because it changes the data boundary.

Design partial-failure behavior. A transcription can succeed while generation times out, and the answer can succeed while synthesis fails. Return the usable text result, permit bounded retry of the failed stage, and avoid rerunning expensive successful work unnecessarily.

15.9 Summary and next steps

ScribeCrate and SpeakCrate provide recorded transcription and speech synthesis for custom applications; `voice-chat` also supplies browser chat. Live captions require ScribeCrate Live, available as an optional definition in the complete platform and `voice-pipeline`. A complete browser voice experience still needs verified UI configuration or companion client logic. Part V adds controlled tools through MCP while keeping tool access distinct from an autonomous agent.

Part V: Tools, Coding Assistants, and AI Agents

Part V exposes selected tools through MCP and connects them to external AI clients. It then uses goose as a practical example of an optional containerized agent that can call the stack's model and tool services without joining the default deployment.

Chapter 16: Expose Controlled Tools with an MCP Gateway

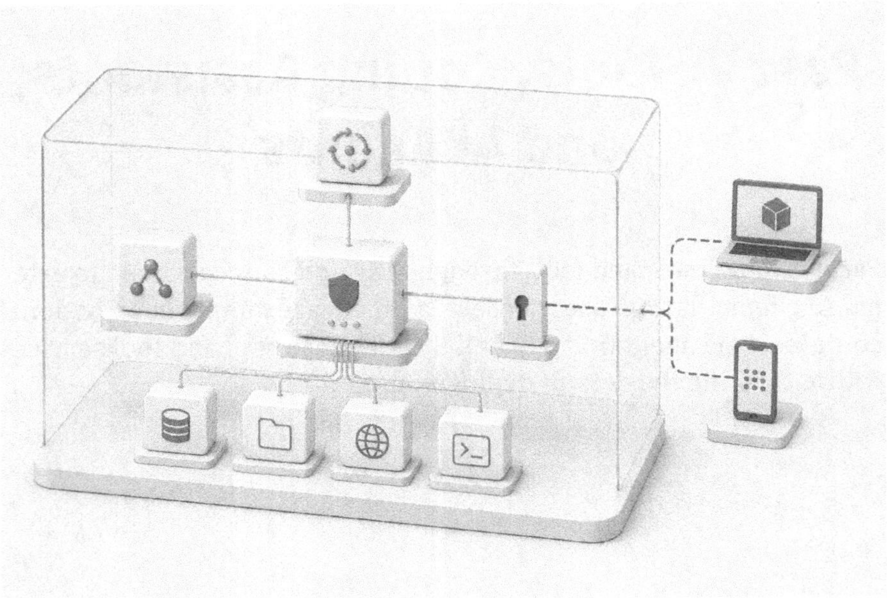

Model Context Protocol, or MCP, gives compatible clients a standard way to discover and call tools. This chapter uses UplinkCrate, the maintained MCPHub and Caddy container that collects selected MCP servers behind one authenticated endpoint. It does not decide goals, run an agent loop, or make a tool safe merely because it uses a standard protocol.

16.1 Understand what MCP provides

An MCP server publishes capabilities such as reading approved files, fetching a URL, querying a database, or working with a source repository. A client presents those tools to a model and mediates calls. The gateway supplies transport, routing, configuration, and authentication for several servers.

Keep three roles distinct:

- The model proposes or selects a tool call.
- The client or agent runtime controls the interaction and approval flow.
- The MCP server executes within the permissions available to its process and credentials.

16.2 Deploy UplinkCrate

The complete platform, `ai-tools`, and `code-assistant` already run UplinkCrate inside their Compose network. Reuse that service and its generated key. Other lightweight variants do not include it; choose a tool variant or plan a separate gateway and explicit client/network connection. The commands below start only the standalone gateway, whose default host port differs from the integrated internal-only access path.

For a fresh standalone gateway, start from the checkout parent:

```
git clone https://github.com/hwdsl2/uplinkcrate
cd uplinkcrate
cp mcp.env.example mcp.env
```

With an existing checkout, enter its current directory and review `mcp.env` for initial servers and external-service credentials. Standalone Compose publishes port `3000` on all host interfaces; for host-only clients, change it to `127.0.0.1:3000:3000/tcp` before startup.

Start the standalone service and inspect its initial logs:

```
docker compose up -d
docker logs --tail=100 mcp
```

The gateway listens on port `3000`. It enables the `fetch` server by default and stores configuration, generated credentials, and first-run state in `mcp-data`. Fresh installs generate both an API key and a dashboard password.

```
docker exec mcp mcp_manage --showkey
docker exec mcp mcp_manage --status
```

In the integrated stack, MCP is internal by default. Publish its port only when a host-side client needs it, and bind it to loopback unless an authenticated TLS proxy is the intended entry point.

16.3 Enable and manage MCP servers

`MCP_SERVERS` selects initial servers. Each enabled capability brings its own configuration and trust boundary:

Server	Required setting	Capability
fetch	None	Fetch URLs and extract content
filesystem	MCP_FILESYSTEM_DIRS	Read or write within mounted directories
github	MCP_GITHUB_TOKEN	Use the GitHub API with the token's permissions
brave-search	MCP_BRAVE_API_KEY	Query the Brave Search API
git	MCP_GIT_REPO	Work with one mounted Git repository
postgres	MCP_POSTGRES_URL	Query a configured PostgreSQL database
memory	None	Use the bundled knowledge-graph memory server
sequential-thinking	None	Use the bundled structured-thinking server

Filesystem and Git paths must exist inside the container, which means the host directories also need explicit bind mounts. Make a mount read-only unless the workflow genuinely needs writes. Use the current MCP environment example (`mcp.env.example`) for the complete first-run settings.

This variable applies when the persistent configuration is first created. Later changes belong in the MCPHub dashboard and survive restarts. Inspect and test configured servers with:

```
docker exec mcp mcp_manage --list
docker exec mcp mcp_manage --test fetch
docker exec mcp mcp_manage --status
```

Avoid deleting the generated settings file merely to reapply an environment list. That resets working configuration and should be treated as an intentional maintenance action.

16.4 Connect UplinkCrate to GatewayCrate

GatewayCrate uses LiteLLM's MCP integration to register the internal gateway as a tool source:

```
mcp_servers:
  - url: http://mcp:3000/mcp
    transport: http
    headers:
      Authorization: "Bearer <mcp-api-key>"
```

Use the internal service address inside Compose. Store the real key through the stack's generated credential mechanism or an uncommitted secret source. Registration makes tools available to a compatible model workflow, but the calling client still needs to decide when and whether calls run. Because client approval and tool-call interfaces vary, use the current documentation for the client you select rather than assuming that one walkthrough applies to every client.

16.5 Connect compatible AI clients

A compatible client needs the streamable HTTP endpoint and Bearer header. The all-server endpoint is `/mcp`; a path such as `/mcp/fetch` narrows routing to one server. Prefer the narrower endpoint when a workflow needs only one capability.

A server-specific endpoint narrows routing; it does not by itself establish an authorization boundary.

Client configuration formats change, so use the maintained UplinkCrate client connection examples for current product-specific fields. Never commit the gateway key in an editor settings repository.

16.6 Test the MCP endpoint

The host-side `127.0.0.1:3000` requests below apply to a standalone gateway whose port you have published. In the complete and tool stacks, MCP stays on the Compose network; run the stack's `stack-check.sh` for its authenticated handshake instead. Do not publish MCP merely to run a local health test.

The unauthenticated health endpoint confirms that the gateway process is reachable:

```
curl -sS http://127.0.0.1:3000/health
```

Then retrieve the key and contact the MCP endpoint:

```
uplink_api_key="$(docker exec mcp mcp_manage --getkey)"
curl -sS http://127.0.0.1:3000/mcp \
  -H "Authorization: Bearer $uplink_api_key"
unset uplink_api_key
```

A protocol-aware client should complete initialization, list the expected tools, and call a harmless operation. A plain `curl` response alone does not prove the full client interaction works.

16.7 Define authentication and permission boundaries

Gateway authentication is one layer of protection. Use the access controls supported by your deployment, and verify both permitted access and expected denials. These controls do not replace restricted mounts, narrowly scoped service credentials, or client approval for consequential actions. Consult the maintained UplinkCrate documentation for supported authentication and permission settings.

- Mount the smallest directory needed, read-only when writes are unnecessary.
- Use repository or database credentials scoped to the exact workload.
- Separate keys between users, environments, and automation where supported.
- Restrict outbound network destinations for fetch and API tools.
- Do not place production secrets in a filesystem tree exposed to a tool.
- Review dashboard changes as security-sensitive configuration.

16.8 Understand the risks of tool-enabled AI

Untrusted web pages, repository files, issues, documents, and database values can contain instructions that attempt to influence the model. Treat such text as data. A client should require approval for writes, external messages, purchases, credential changes, destructive operations, and other consequential actions.

Log the requesting identity, selected tool, bounded arguments, outcome, and approval decision without recording unnecessary secrets. Tool output must not automatically become trusted instructions for the next call.

16.9 Secure and troubleshoot UplinkCrate

If a server is missing, confirm first-run versus dashboard configuration and required credentials. If a client cannot connect, check its transport, URL, Bearer header, TLS, and proxy buffering. For filesystem or Git failures, inspect the container mount and operating-system permissions.

Verify that protected MCP and administrative routes reject requests without valid credentials. The public `/health` endpoint confirms reachability, not authorized tool access. Keep gateway credentials and dashboard passwords in the backup plan, use HTTPS outside a trusted host, and rate-limit public access.

16.10 Summary and next steps

UplinkCrate now exposes selected tools through an authenticated, observable boundary. Chapter 17 combines that boundary with local models and, when needed, embeddings in two stacks designed for external clients.

Chapter 17: Build a Private Code Assistant and AI Tools Stack

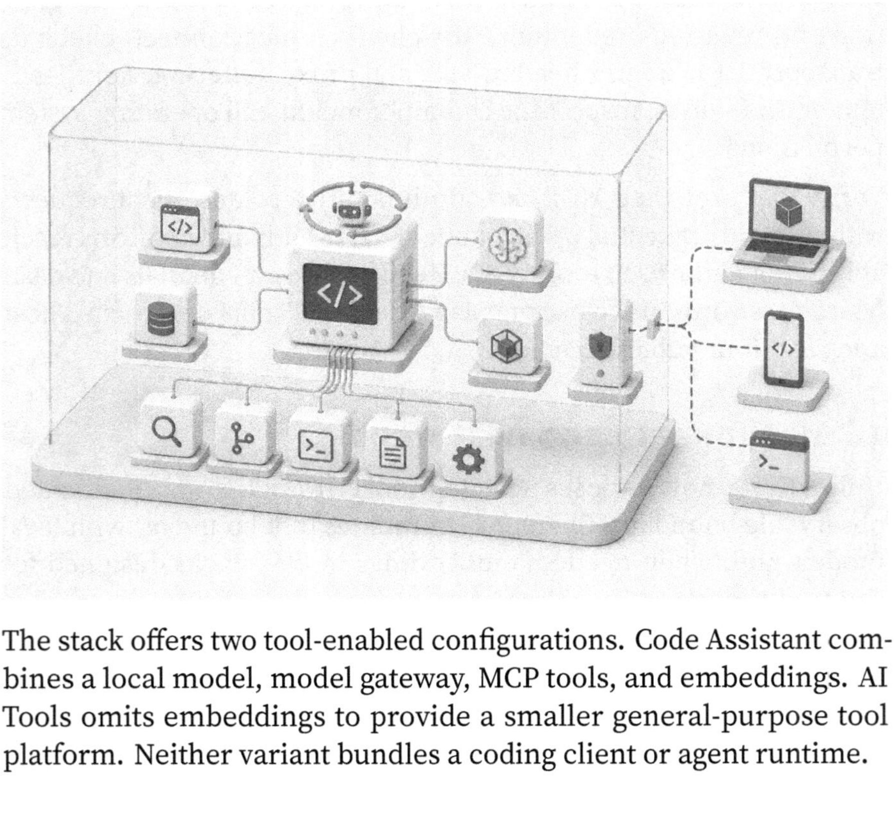

The stack offers two tool-enabled configurations. Code Assistant combines a local model, model gateway, MCP tools, and embeddings. AI Tools omits embeddings to provide a smaller general-purpose tool platform. Neither variant bundles a coding client or agent runtime.

17.1 Compare the code assistant and AI tools stacks

Code Assistant uses InferCrate, GatewayCrate, UplinkCrate, and EmbedCrate. Its documented memory estimate is about 5 GB with a 3B model. Choose it when the intended client needs semantic code search or another vector workflow.

AI Tools uses InferCrate, GatewayCrate, and UplinkCrate. It is the simpler choice when the client needs model and tool endpoints but not an embeddings service. Both depend on an external compatible client to manage conversations, approvals, and file changes.

The complete platform already runs these services, including EmbedCrate. Reuse it if you followed Chapter 3; the variant blocks below are smaller alternative starts. If another lightweight variant is running, stop it without deleting volumes before choosing either tool variant. Do not run unmodified variants side by side.

Both publish GatewayCrate's HTTP port `4000` on all host interfaces by default; restrict it to the host or a trusted network before remote access.

Both variants are maintained in the companion repository. From the directory where you keep project checkouts, clone it before choosing a variant:

```
git clone https://github.com/hwdsl2/self-hosted-ai-stack
```

If you already have the repository, reuse that checkout. Each alternative deployment block below assumes your shell starts in the directory that contains the `self-hosted-ai-stack` checkout.

17.2 Deploy the code assistant configuration

```
cd self-hosted-ai-stack/stacks/code-assistant
docker compose up -d
docker exec ollama ollama_manage --pull llama3.2:3b
../../stack-check.sh
```

GatewayCrate is published on port `4000`, EmbedCrate is on host loopback port `8000`, and MCP is internal unless its port mapping is enabled for a host-side client. Use the CUDA file on a supported NVIDIA host.

17.3 Connect models, embeddings, and MCP tools

Each endpoint has a separate responsibility and credential:

- GatewayCrate provides chat or completion access to the configured local model.
- EmbedCrate converts text into vectors but does not store a code index.
- UplinkCrate exposes the specific file, Git, fetch, or remote tools you enabled.

Do not give a client a single broad credential when the services can remain separated. Host-side clients need only the ports they actually call; containerized clients can join a controlled network.

17.4 Connect a compatible coding client

Configure the client with an OpenAI-compatible base URL for GatewayCrate and a model alias that exists in the gateway. Configure MCP separately with the streamable HTTP URL and Bearer token. Client configuration fields vary, so verify them against the version you install.

Start with read-only repository access and a disposable branch. Confirm that the client can explain a file and propose a patch before permitting shell commands, commits, pushes, or external services.

The core chapter intentionally uses a client-neutral connection contract. Named clients change configuration formats and supported transports quickly. Use the client's maintained documentation for version-specific setup.

17.5 Add semantic code search

The embeddings service can represent functions, files, documentation sections, or other code chunks. The client still needs to parse content, choose boundaries, store vectors, filter by repository and revision, and update the index when code changes. Reuse the pgvector pattern from Part III if durable storage is needed.

Embedding search is complementary to exact symbol, filename, and text search. Code identifiers often benefit from exact search, while semantic retrieval can help with architectural questions or concepts expressed in different words.

17.6 Deploy the lighter AI tools stack

If you started the code assistant variant, stop it from that stack directory before switching:

```
docker compose down
cd ../../..
```

The second command returns to the directory that contains the repository checkout. Run the lighter alternative from there:

```
cd self-hosted-ai-stack/stacks/ai-tools
docker compose up -d
docker exec ollama ollama_manage --pull llama3.2:3b
../../stack-check.sh
```

Use this variant when a deterministic application or external client needs a local model plus a controlled tool gateway. Stop another lightweight variant first because defaults share container and volume names.

17.7 Select tools for a workflow

Enable tools from the required outcome backward:

- For repository explanation, start with read-only filesystem and Git status or diff.
- For issue triage, use a narrowly scoped repository token.
- For documentation lookup, restrict fetch destinations when possible.
- For database analysis, use a read-only role and approved schemas.
- For editing, mount only the target worktree and require review before commit.

Avoid enabling every available server for convenience. A smaller tool set reduces mistaken selection, credential exposure, and the amount of untrusted output entering the model context.

17.8 Secure and troubleshoot the stacks

Keep model and tool endpoints on loopback or internal networks, protect public entry points with TLS, and separate repository credentials from model keys. Do not mount a home directory, container socket, or broad secrets directory into an MCP service.

Diagnose one boundary at a time: model directly through GatewayCrate, embeddings directly, MCP health and tool listing, then the client. Verify the client is using the expected URL and model rather than a silent hosted fallback.

17.9 Summary and next steps

A tool variant supplies reusable local model and controlled-tool endpoints for external clients. Code Assistant also supplies EmbedCrate for client-built semantic search. Chapter 18 explains when to use native goose, gives a bounded container example with the AI Tools variant, and links to its full maintained procedure. It also sets explicit permissions, approvals, and stop conditions.

Chapter 18: Connect and Control a Self-Hosted AI Agent

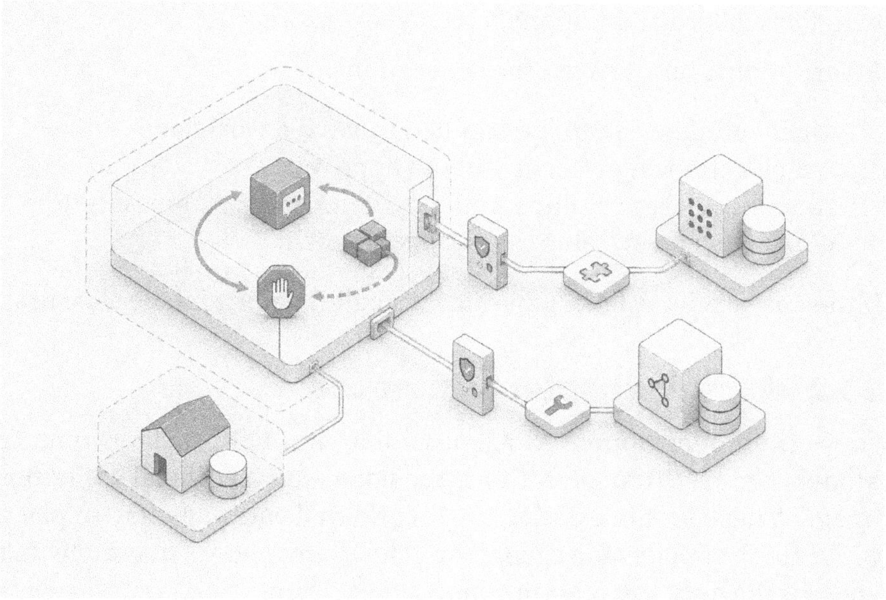

Self-Hosted AI Stack supplies model and tool gateways, but does not bundle or install an agent runtime. For ordinary work with project files, install goose on your workstation and connect it to the stack. This chapter also shows a bounded task in a temporary goose container with separate state and no initial access to host files; the maintained guide supplies its full setup.

The optional container uses GatewayCrate and UplinkCrate over the internal Docker network. Its Compose profile excludes it from ordinary stack startup, so you launch it only when you need that execution boundary.

18.1 Distinguish tools, workflows, and agents

A tool performs a bounded operation. A deterministic workflow connects known steps with explicit branching and error handling. An artificial intelligence (AI) agent lets a model select and repeat steps while pursuing an objective.

Prefer the least autonomous mechanism that solves the problem. A scheduled transcription and summary pipeline does not need an agent. A support workflow with known retrieval and ticket steps may need conditional code but not open-ended planning. An agent becomes useful when the route cannot be specified reliably in advance and the benefit exceeds the additional risk and operating cost.

Before adding one, answer these questions:

1. What outcome cannot be handled by a fixed workflow?
2. Which actions may occur without approval?
3. How will success, failure, and maximum effort be measured?
4. What data and systems can the agent reach?

If the answers are vague, automate a smaller deterministic slice first.

18.2 Why this chapter uses goose

goose is an open source AI agent with a LiteLLM provider, remote Model Context Protocol (MCP) extensions, and an official container image. These let the existing model and tool gateways stay in place while goose supplies the agent loop. The example does not imply that goose is the only suitable runtime.

The durable design is the separation among agent, model gateway, and tool gateway. Image versions, CLI flags, and configuration menus change; the maintained guide and current goose documentation carry the execution details.

18.3 Choose native goose for workstation workflows

Use native goose for routine interactive work with project files on your workstation. It can use the operating-system keyring and local approval prompts, but it has your account's file access. Start in a dedicated project. All maintained variants supply GatewayCrate; the complete and tool variants also supply UplinkCrate when the agent needs controlled tools.

The maintained guide at **selfhostedaistack.com/goose** has separate workstation and container paths. It covers Linux, macOS, and Windows setup, restricted credentials, model guidance, permissions, and testing. The stack can remain on a Linux server while native goose runs on your workstation.

Use the container path when you need a restricted, reproducible server-side environment with explicit mounts and separate agent state.

18.4 Understand the containerized architecture

The existing platform supplies model and controlled-tool interfaces. The goose container adds the planning loop, session state, and approval experience.

```
User -> temporary goose container -> LiteLLM -> Ollama or
  hosted model
                         |
                         +-> MCP gateway -> enabled MCP servers
                         |
                         +-> goose home volume
```

goose, GatewayCrate, and UplinkCrate share the Compose network. goose calls the gateways by internal service name, so neither the agent nor UplinkCrate needs a new published port for this example. The maintained guide gives the current service addresses.

A prompt can still cross external boundaries. If GatewayCrate routes to InferCrate, the model request remains on infrastructure you operate. If the selected alias routes to a hosted provider, that provider receives the prompt and supplied context. An MCP server can also contact an external service. Containerization isolates files and processes, but it does not make outbound traffic local.

18.5 Prepare the optional container path

In the maintained guide linked above, choose the container path using the AI Tools variant from Chapter 17. Reuse AI Tools if it is already running. If you chose the complete platform or another lightweight variant, do not start AI Tools beside it with the default names and volumes. The container procedure has not been verified as an overlay for every variant. Use native goose with the current stack, or back up and stop it before switching to AI Tools for the container procedure.

The container guide supplies the complete optional Compose overlay. Its profile keeps goose out of ordinary startup. The procedure checks the image version and records its digest, gives the agent separate persistent state, and uses an expiring GatewayCrate key restricted to one model alias plus a separate UplinkCrate key. The initial configuration publishes no port and mounts no host workspace or Docker socket. It drops Linux capabilities and disallows privilege escalation.

The named home volume still stores configuration, extension credentials, sessions, and logs. Users with Docker administration access can inspect a running container's environment. Treat the volume as sensitive and keep both keys narrow; do not store them in a tracked environment file. Choose a model alias that supports tool calling. A small model can confirm connectivity without being reliable at multi-step tool use.

18.6 Try one bounded read-only task

Follow the maintained container path to start AI Tools, create the overlay and restricted keys, inspect the current goose image and CLI, and configure only one authenticated fetch extension. Confirm that the effective provider is LiteLLM, the selected model is the intended alias, and unneeded extensions, including shell and file tools, are disabled. If the selected image lacks the expected session limits or you cannot restrict its tools, stop and reconcile its documentation before running the example.

After those checks, use the guide's current command to launch a temporary session with small limits on turns and repeated tool calls. Confirm that the selected image supports the limits before starting. This example was checked against source and Compose configuration, not by an end-to-end container run. The guide maintains the version-specific command and compatibility checks.

Give the agent a narrow request:

> Use only the approved fetch extension. Retrieve https://example.com, report the page title, and summarize the page's purpose in two sentences. Do not use shell or file tools.

The prompt describes intent, but it is not the security boundary. The absent host mounts, disabled broad extensions, limited MCP server set, separate credentials, approval mode, dropped capabilities, and command-line limits provide the meaningful controls.

Before approving a call, check the extension and URL. Afterward, verify that goose used the chosen GatewayCrate alias, only the fetch tool was available, GatewayCrate and UplinkCrate recorded the requests without exposing secrets, and the temporary container was removed. A fluent answer alone does not prove the intended path worked. Treat a mismatch as a failed boundary check and stop. The online procedure includes the exact setup, compatibility checks, log inspection, and sanitized troubleshooting steps.

18.7 Understand what the container boundary does

The container prevents goose from seeing arbitrary host files unless you mount them. It also gives the agent its own process, filesystem, and Linux capability boundary. Those protections are valuable, but incomplete:

- The agent can reach services allowed by its Docker network and outbound network policy.
- It can modify its writable home volume.
- Docker administrators can inspect or control the container.
- A mounted directory grants the container the access specified by that mount.
- Prompt injection can still influence model-selected tool calls.
- Approval can still be mistaken or too broad.

Never mount `/var/run/docker.sock` into an agent container. Access to the Docker socket commonly provides control over the host and the other containers, which would defeat the isolation this design is intended to provide.

18.8 Add filesystem access only when required

The fetch example needs no workspace. If a later task must inspect files, add one explicit mount to the goose service instead of mounting a home directory. Start with read-only access to a dedicated project path; the maintained guide shows the current Compose syntax.

Use a dedicated path containing only material the agent may read. If the task must write, use a disposable worktree or staging directory, change only that mount to read-write, and require approval before any commit, publication, external message, or production action.

Separate read, propose, and execute permissions. The container may prepare a change in a dedicated worktree while a human-controlled process performs the final commit, push, deployment, or other consequential action.

18.9 Apply tool permissions and stop conditions

Classify each enabled tool by the authority the task needs:

- Allow automatically only operations that are truly safe and read-only in the current environment.
- Require approval for writes, commands, external side effects, or operations whose target can vary.
- Deny capabilities the task does not require.

The maintained guide maps these choices to the current goose permission settings.

Human confirmation should normally gate deletion, publication, external messages, purchases, production changes, credential operations, privilege changes, and irreversible or high-impact writes. Approval belongs immediately before the consequential action, after the exact target and payload are known.

Set limits that correspond to the potential impact of the task:

- Maximum turns and repeated identical tool calls
- Maximum elapsed time and retries
- Provider cost or token budget where applicable
- Allowed directories, hosts, repositories, and database roles
- A stop on authentication failure, ambiguous targets, or repeated tool errors
- A stop when the task needs authority outside the original request

Record the task identity, image digest, selected model, enabled tools, sanitized arguments, approvals, results, and final state. Keep enough information to reconstruct a failure without turning logs into another store of secrets or personal data.

18.10 Test failure and abuse cases

A successful fetch proves only the basic connection. Before permitting writes or production access, test:

- Web pages or documents containing prompt injection
- Tool output that falsely claims an action succeeded
- Timeouts after a partial write
- Duplicate retries and repeated identical calls
- Missing or expired credentials
- Permission denial and unavailable approval
- Attempts to reach unapproved networks, files, or environment variables
- Results that exceed size, turn, time, or cost limits
- Attempts to access host paths or the Docker socket

Release a workflow only after these tests show that authorization, approvals, idempotency, budgets, and stop conditions behave as intended.

18.11 Do not make the first agent an unattended service

An unattended agent is not the starting pattern for this book. It needs an authenticated user interface or API, concurrency policy, approval channel, scheduler, audit trail, cancellation behavior, and recovery design.

Keep the initial agent opt-in and temporary. Add unattended execution only for a narrowly defined task after its credentials, approval behavior, failure recovery, and stop conditions have been tested independently. The guide gives the current launch and removal commands.

18.12 Troubleshoot by boundary

When a session fails, locate the failing boundary before changing settings: Compose and required variables for startup, GatewayCrate alias and restricted key for model requests, then MCP endpoint, Bearer header, and enabled server for tools. Stop immediately if unexpected tools appear. The maintained guide has the corresponding commands;

avoid verbose configuration or debug output that could reveal credentials, prompts, or returned documents.

18.13 Remove the optional agent

Exit the session and confirm its temporary container is gone. Revoke the GatewayCrate virtual key, remove the Compose overlay, clear shell credentials, and review the agent home volume under your backup and retention policy. Removing the overlay does not delete the named volume or its session and credential data. Removing that volume does not delete Ollama models or gateway state. The online procedure gives the exact removal sequence.

18.14 Know when an agent is unnecessary

Do not keep the agent merely because the integration works. A fixed script or workflow is often easier to test, reproduce, and secure. Use an agent only when model-selected iteration produces a meaningful improvement and the task can be bounded by permissions, approvals, and measurable stop conditions.

18.15 Summary and next steps

You now have a bounded pattern for connecting a temporary goose container to restricted model and tool endpoints, with separate state and no initial host workspace mount. Use the maintained guide for the current container commands; for routine workstation work, choose its native goose path. Before granting filesystem writes, broader tools, or unattended operation, verify every boundary with the exact image you selected. Part VI applies the same discipline across the platform.

Part VI: Secure and Operate the Stack

Part VI turns the deployment into a system you can operate responsibly. You will customize it without breaking its internal wiring, minimize exposure, diagnose failures by dependency, and prepare tested recovery, migration, and upgrade paths.

Chapter 19: Customize the Deployment

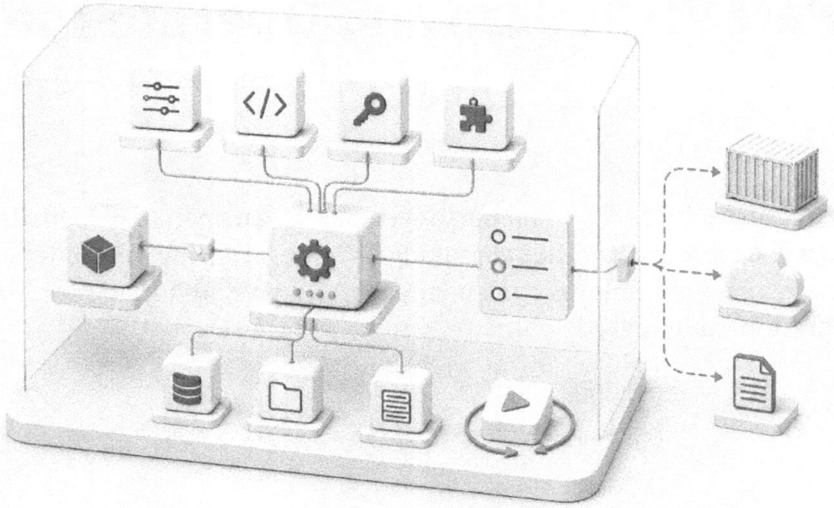

Customize from a working configuration, make one bounded change at a time, and preserve the generated credential and data wiring. This method keeps your deployment maintainable as the source stack evolves.

This chapter assumes you have a working checkout of the companion repository from Chapter 3 or a later workflow chapter. Run Compose commands from the repository root for the complete stack or from the selected directory under `stacks/` for a lightweight variant.

19.1 Configure services with environment files

Each component provides an example environment file. Copy only the examples for services that need overrides, keep the real files out of version control, and mount them through the documented Compose entry. An unset value can trigger secure generated behavior, while an explicitly empty value can have a different meaning. Follow the stack customization documentation and verify the relevant example environment file before assuming that unset and explicitly empty values are equivalent.

The component chapters include curated tables for settings that materially change behavior, capacity, security, integration, or offline operation. They are decision aids, not exhaustive references. The maintained environment example remains the authority for every supported variable and its current default. This avoids copying long catalogs into the book while still making the controls that shape the architecture readily discoverable.

Record why every override exists. Prefer stable service names and internal ports so containers can communicate even when host-facing ports change.

19.2 Use Docker Compose overlays

The repository provides separate CPU and CUDA Compose files plus a proxy overlay. File order matters because later definitions merge with or replace earlier values.

```
docker compose -f docker-compose.yml config
docker compose -f docker-compose.cuda.yml config
docker compose -f docker-compose.yml -f docker-compose.proxy.
  yml config
```

Inspect rendered images, ports, networks, volumes, environment, and GPU reservations before starting. Use the same file set for `up`, `down`, `pull`, `logs`, and `config`.

19.3 Run individual containers

The stack's standalone-container instructions provide supported `docker run` paths. This can reduce the number of services, but it transfers dependency ordering, networks, credentials, volumes, and update commands to the operator. Prefer Compose when several services must share generated keys or database settings.

19.4 Use Podman

The root stack's Podman instructions describe support on a best-effort basis. Consult them for current version requirements before deployment. The CPU Compose path is intended to work as written, while GPU and SELinux hosts need additional steps. Compose GPU reservations do not apply directly to Podman, so NVIDIA access uses the documented CDI path. Confirm networking, rootless-port behavior, volume labels, health checks, and command compatibility on the actual host.

19.5 Change ports, models, and providers

A host port change affects clients and proxies, not necessarily container-to-container URLs. A container port change affects health checks, service discovery, and dependent configuration. Update all consumers deliberately.

Model changes affect memory, startup time, aliases, context limits, and sometimes stored vectors. Provider changes can move prompts and retrieved content outside local infrastructure. Keep application code on stable GatewayCrate aliases and document the data boundary behind each alias.

19.6 Manage CPU and CUDA variants

CUDA usually changes both image tag and device reservation. Confirm every enabled inference service uses the intended variant and that optional services are not left on CPU accidentally. CUDA images in these projects are generally `linux/amd64`; CPU images support broader architectures.

Do not combine CPU and CUDA files casually. Render the selected file and inspect it before deployment.

19.7 Add or remove services safely

Map dependencies first. GatewayCrate needs PostgreSQL and model providers; AnythingLLM needs GatewayCrate; optional clients may need generated key-sharing volumes. When adding a service, define its network, persistent data, authentication, readiness, resource limits, and backup policy. When removing one, retain its volume until recovery is no longer required.

19.8 Validate a customized deployment

Use this sequence:

1. Render the effective Compose configuration.
2. Confirm images, bind addresses, volumes, and dependencies.
3. Start the stack and wait for readiness.
4. Run the repository's health-check script (stack-check.sh) from the appropriate location.
5. Test the customized request path, not only generic health endpoints.
6. Restart once and confirm credentials and data persist.
7. Record the working source commit and local overrides.

19.9 Summary and next steps

Customization is safest when treated as a small, reviewable layer over a verified source configuration. Chapter 20 applies the same discipline to exposure, credentials, administrative interfaces, and tool permissions.

Chapter 20: Secure Local and Internet-Facing Services

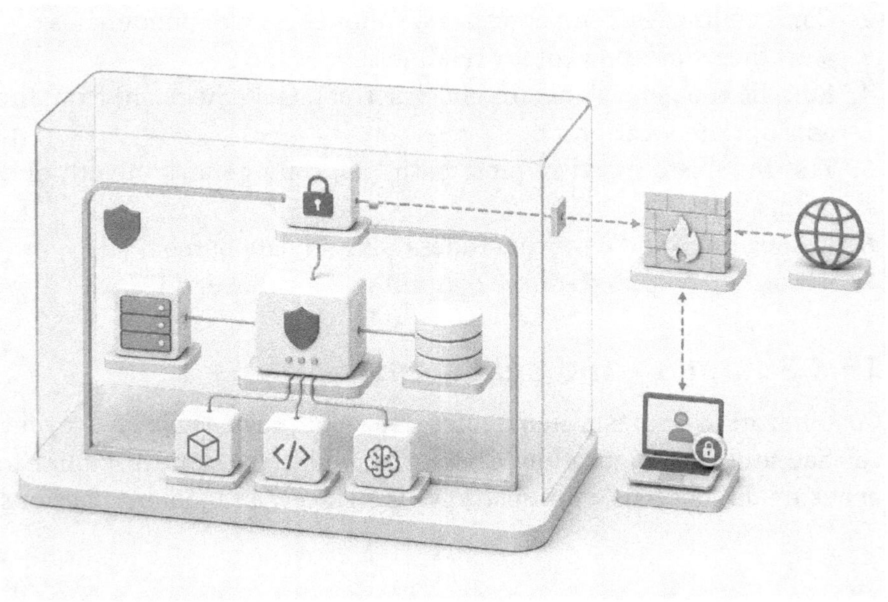

Self-hosting changes who operates the security boundary. It does not make a service private by itself. This chapter reduces the reachable surface, protects credentials, and makes local and external data flows explicit.

The executable service bindings, generated-credential wiring, and proxy configuration used throughout this chapter make those controls concrete.

20.1 Start with a threat model

List users, administrators, client applications, model providers, stored data, uploaded content, networks, credentials, and enabled tools. Identify which components accept untrusted input and which can change external state. Then define what must remain confidential, available, and recoverable.

A single-user lab, a trusted team network, and a public service need different controls. Revisit the model when a port, provider, user population, or tool permission changes.

20.2 Manage generated API keys and credentials

Fresh persistent installations generate keys for major stack services. Retrieve them through their management commands, keep them out of shell history and repositories, and store them in a password or secret manager. After deployment or an upgrade, verify that protected endpoints reject requests without valid credentials.

Rotate a key after exposure and update all dependent clients. Back up key-bearing volumes because regeneration can break clients and shared service wiring. Use separate, restricted database and remote provider credentials where the current stack permits it.

20.3 Bind services to localhost

A binding such as `127.0.0.1:9000:9000` is reachable only from the host, while `9000:9000` normally publishes on all host interfaces. Inspect effective mappings with:

```
docker compose ps
docker compose config
```

Keep PostgreSQL, InferCrate, MCP, and component APIs internal unless a specific host client needs them. Publishing an authenticated port still increases attack surface.

20.4 Add HTTPS with the Caddy overlay

The included Caddy proxy overlay (docker-compose.proxy.yml) makes Caddy the public listener and rebinds selected direct ports to loopback. It requires current Docker Compose, valid DNS records, reachable ports 80 and 443, and no conflicting listener.

```
DOMAIN=chat.example.com ACME_EMAIL=admin@example.com \
  docker compose -f docker-compose.yml -f docker-compose.
  proxy.yml up -d
```

Confirm both HTTPS access and the absence of direct public access. Back up `caddy-data` because it contains private keys and ACME state.

20.5 Use another reverse proxy

An alternative proxy must terminate TLS, preserve required authorization headers, support request and response streaming, upgrade WebSockets where needed, impose body and time limits, and forward to a loopback or internal address. Configure origin and CORS behavior only for intended browser clients.

ScribeCrate Live non-browser clients can authenticate the WebSocket handshake with an `Authorization` header. Browser clients use a `token` query parameter because their WebSocket API cannot add arbitrary headers. Redact query strings containing credentials from proxy access logs, use WSS, and keep the direct service port on loopback as defense in depth.

20.6 Protect administrative interfaces

AnythingLLM has its own password, and GatewayCrate uses its master key for administrative access. MCPHub configuration controls executable tools. Treat these interfaces as privileged even if the related API uses another credential. Restrict them by network, strong authentication, and preferably a separate administrative access path.

Do not expose PostgreSQL directly for convenience. Use a restricted tunnel or local administrative session when access is necessary.

20.7 Secure MCP tools

Gateway authentication is only the first layer. Use read-only mounts, scoped tokens, restricted database roles, outbound network policy, client approvals, and an audit log. Treat tool content as untrusted, and test prompt injection from fetched pages, issues, documents, and repository files.

20.8 Minimize exposed services

For each published port, record the owner, audience, authentication, TLS path, rate limit, and reason it cannot remain internal. Remove mappings that have no current consumer. Scan from another machine, not only from the server itself, to verify the real boundary.

Appendix A summarizes the maintained exposure model. Before exposing a stack, render the effective Compose configuration for the selected CPU, CUDA, and proxy files and verify every published port. A single changed mapping can alter the public boundary.

20.9 Understand external provider data flow

GatewayCrate can route to local InferCrate or hosted providers. A hosted route may receive system instructions, user prompts, conversation history, retrieved document text, code, and tool results. The fact that the gateway runs locally does not keep that payload local. Label model aliases and user interfaces so operators know which boundary each request uses.

20.10 Security review checklist

- Every published service has an owner and documented need.
- Direct HTTP ports are internal or on loopback behind TLS.
- Authentication is verified, including upgraded installations.
- Admin interfaces are restricted more tightly than ordinary use.
- Upload size, rate, concurrency, and timeout limits are set.
- Tool mounts, tokens, database roles, and outbound access follow least privilege.
- Hosted model routes and retained data are disclosed.
- Backups are encrypted, access-controlled, and restorable.
- Logs avoid unnecessary prompts, transcripts, documents, keys, and tokens.
- Security and dependency updates have a tested deployment path.

20.11 Summary and next steps

A protected deployment has a small reachable surface, explicit access controls, bounded inputs, constrained tools, and known data flows. Chapter 21 uses that map to diagnose failures without making destructive changes.

Chapter 21: Diagnose Problems and Verify Stack Health

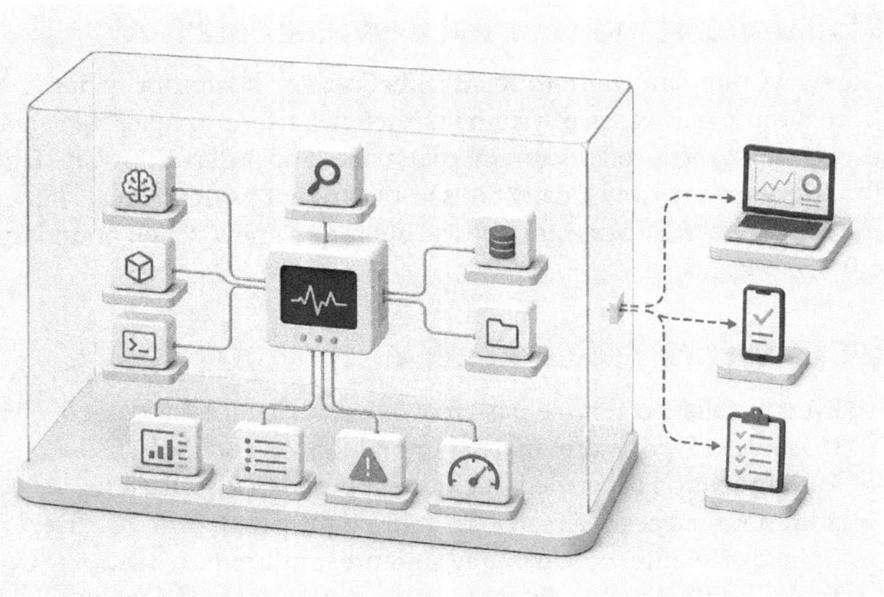

Reliable troubleshooting follows dependencies from the earliest failing boundary. Avoid deleting volumes, regenerating configuration, or changing several settings before collecting evidence.

The maintained Self-Hosted AI Stack troubleshooting guide (docs/troubleshooting.md) contains the current stack-specific commands, failure signatures, and issue-reporting guidance that accompany this chapter's diagnostic method.

21.1 Follow a repeatable triage workflow

1. Define the failing request, expected result, and last known working state.
2. Record recent image, configuration, model, network, and host changes.
3. Render the effective Compose configuration.

4. Inspect container state and readiness.
5. Test dependencies in order.
6. Read bounded logs around the failure.
7. Change one cause at a time and repeat the exact request.

21.2 Inspect container state, health checks, and logs

```
docker compose ps
docker compose logs --tail=100
./stack-check.sh
```

If the stack was started with overlays, repeat the same file selection during diagnosis so Compose evaluates the same project:

```
docker compose -f docker-compose.yml \
  -f docker-compose.proxy.yml ps
docker compose -f docker-compose.cuda.yml \
  -f docker-compose.proxy.yml logs --tail=100 litellm
```

`stack-check.sh` detects Podman automatically. Set `CONTAINER_ENGINE=podman` when you need to select it explicitly.

For a sub-stack, run its relative `../../stack-check.sh`. A running container is not necessarily a ready application. First model downloads, database migrations, and dependency waits can delay useful traffic.

21.3 Diagnose startup and readiness failures

Check the earliest unhealthy dependency. PostgreSQL and InferCrate precede GatewayCrate, and UplinkCrate also precedes it in variants that include tools. GatewayCrate precedes AnythingLLM. Optional services may load models after their processes start. Inspect health endpoints and service-specific readiness behavior instead of repeatedly restarting the whole stack.

21.4 Diagnose Ollama and model problems

```
docker exec ollama ollama_manage --listmodels
docker compose logs --tail=100 ollama
```

Confirm the requested model was pulled, the alias matches, memory is sufficient, the API key wiring is intact, and clients inside Compose use `http://ollama:11434` rather than host loopback.

21.5 Diagnose LiteLLM and provider errors

```
curl -sS http://127.0.0.1:4000/health/liveliness
docker exec litellm litellm_manage --showkey
docker compose logs --tail=100 litellm
```

Separate database startup, gateway authentication, missing model alias, local-provider connectivity, and hosted-provider credential or quota errors. Send one direct request through the gateway before debugging a user interface.

21.6 Diagnose MCP connectivity problems

```
docker exec mcp curl -sf http://127.0.0.1:3000/health
docker exec mcp mcp_manage --list
docker exec mcp mcp_manage --status
```

Confirm the client transport and Bearer header, then test one server. For host-side clients, verify the port is intentionally published. For server failures, inspect mounts, credentials, network access, and dashboard configuration.

21.7 Diagnose optional service failures

Test each API directly with a small known input and the service's own key. For ParseCrate, distinguish liveness from readiness. For EmbedCrate and ScribeCrate, distinguish the active server model from the compatibility model named in a request. For SpeakCrate, distinguish binary audio output from a JSON error. Verify GatewayCrate before changing AnythingLLM settings.

21.8 Diagnose GPU and CUDA failures

Confirm the host driver and NVIDIA Container Toolkit work, the CUDA Compose file was used, the image supports the architecture, and the container can see the GPU. Only then inspect model-specific compute settings. A healthy CPU fallback can hide a GPU configuration error, so verify actual device use.

21.9 Diagnose reverse proxy and networking problems

Check DNS, public firewall rules, port conflicts, certificate logs, direct bind addresses, upstream service health, and proxy timeouts. Streaming endpoints may need buffering disabled or longer timeouts. WebSockets also need an upgrade path and origin policy.

For the verified companion ScribeCrate Live image, a missing or incorrect WebSocket key should fail the handshake with HTTP 401. An HTTP 500 during an authenticated handshake can indicate a server-side authentication error or an incompatible WhisperLive and `websockets` combination. Confirm the image revision, then test a valid Bearer header and a valid `token` query parameter before debugging audio framing.

21.10 Prepare a useful issue report

Include host OS and architecture, Docker and Compose versions, source commit, selected stack and Compose files, image tags, exact reproduction steps, expected and actual results, container state, health output, and bounded logs. Redact keys, database URLs, provider tokens, private hostnames, documents, prompts, transcripts, and personal data.

Do not include an entire environment file. Replace secret values while preserving whether each setting was set, empty, or absent.

21.11 Summary and next steps

Dependency-first diagnosis produces evidence without destroying the state needed to understand the failure. Chapter 22 protects that state and provides recovery, migration, and upgrade practices.

Chapter 22: Back Up, Restore, Upgrade, and Migrate

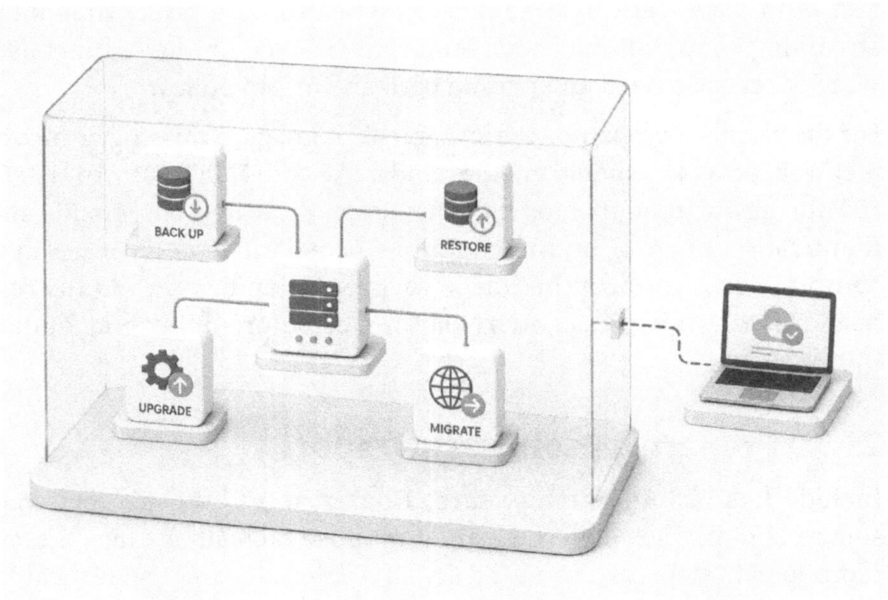

A backup is useful only when it covers the required state, is protected from the same failure as the host, and can be restored. Define acceptable data loss and recovery time before choosing a procedure.

The maintained Self-Hosted AI Stack backup and restore guide (docs /backup-restore.md) contains the current executable backup, restore, migration, and pre-upgrade procedures. This chapter explains how to choose and operate those procedures safely without duplicating a volume list that can change with the stack.

22.1 Understand persistent volumes

The stack uses named volumes for models, generated keys, Gateway-Crate and PostgreSQL data, MCP settings, AnythingLLM workspaces and authentication, optional-service caches, and Caddy certificate state. Shared key-transfer volumes are regenerated and generally do not need backup. Their authoritative key-bearing volumes do need protection.

Classify data before optimizing backup size:

- Irreplaceable or user-owned: chat workspaces, uploaded sources, custom models, application data.
- Security and configuration state: API keys, database data, MCP settings, passwords, TLS private keys.
- Regenerable but expensive: downloaded models and conversion caches.

Appendix A summarizes the volume roles. Check the maintained backup guide and the effective Compose configuration before each backup because the volume set can change as services evolve.

22.2 Export generated credentials

Use each service's machine-readable management command and store the result in an encrypted secret system. Do not print a combined key file to a shared terminal or commit it. Preserve `ai-stack-shared` with PostgreSQL because fresh Compose deployments keep the generated database password there.

22.3 Choose a cold or hot backup

A cold backup stops the stack before archiving volumes and provides the simplest consistency model. Use it for complete recovery points and migrations when downtime is acceptable.

PostgreSQL supports a consistent hot backup through `pg_dump`. Model cache volumes are usually stable after downloads complete, but AnythingLLM, MCP, GatewayCrate configuration, and Caddy state can change. Follow the linked guide's service-specific downtime table rather than declaring every live archive safe.

22.4 Back up all volumes

The linked backup and restore guide contains the authoritative loop for the current volume set. In outline:

1. Export credentials securely.

2. Stop the selected stack for a cold backup.
3. Create one archive per existing named volume through a temporary container.
4. Record source commit, image digests, Compose files, checksums, and backup time.
5. Encrypt and copy archives to separate storage.
6. Restart and run the stack health check.

Do not copy raw PostgreSQL files from a running database. Use a cold volume archive or `pg_dump`. Use the exact archive loop in the maintained backup guide so its volume list and safety checks stay aligned with the selected stack revision.

22.5 Back up and restore a single service

Single-volume recovery is useful for a damaged model cache or component configuration, but services can share dependencies and keys. Stop writers, archive or preserve the current target, restore into a named volume, start dependencies first, and test both the service and its consumers.

Restoring is destructive to the target volume. Resolve its exact name and keep the current data until the recovered service has been validated.

22.6 Restore the complete stack

On a prepared host, obtain the recorded source version, render the same Compose variant, create or restore named volumes, and start the stack. Verify generated keys, PostgreSQL contents, model listings, AnythingLLM login and workspaces, MCP servers, optional APIs, and TLS behavior.

Do not claim success because containers are running. Exercise one real request through every critical path.

22.7 Migrate to another server

Check CPU architecture, GPU support, filesystem capacity, Docker and Compose versions, DNS, firewall, and image availability on the target. Transfer encrypted archives and checksums separately from decryption material. Restore without exposing services publicly, validate locally, then change DNS or client endpoints.

Downloaded models may be portable between compatible images and architectures, but validate them. A clean re-download can be safer than carrying a stale cache when bandwidth permits.

22.8 Prepare for an upgrade

Read the stack changelog (CHANGELOG.md) and applicable component release notes, record current images and configuration, export keys, take a tested recovery point, and render the new Compose configuration. Pull and recreate using the same selected files, wait for migrations and model readiness, then run health and functional tests.

Avoid floating upgrades without a rollback reference. Record image digests or working version tags.

22.9 Roll back a failed upgrade

Stopping new containers and selecting an older image does not guarantee database or configuration compatibility. If an upgrade migrated data, restore the matching pre-upgrade volumes as a unit. Keep the failed state for diagnosis when storage permits, and never overwrite the only recovery point.

22.10 Verify recovery

- Keys match the clients that should continue working.
- PostgreSQL and vector records are present and queryable.
- Ollama models load and route through GatewayCrate.
- AnythingLLM authentication, workspaces, and documents persist.
- MCP settings, mounts, and restricted credentials are correct.
- Optional document and speech APIs complete known requests.
- Proxy certificates and direct bind restrictions are correct.
- A second backup succeeds from the restored deployment.

22.11 Summary and next steps

The platform now has an operational lifecycle: customize carefully, expose minimally, diagnose by dependency, and recover from tested backups. The conclusion brings these practices together and helps you choose the next bounded use case to build.

Conclusion: Where to Go Next

You do not need every service in this book to gain value from self-hosting. A useful platform begins with one bounded workload, an understood data path, and an operator who can recover it. The larger architecture is a menu of compatible capabilities, not a requirement to run every container.

Choose your next use case

Choose one useful outcome and operate it well. That may be private chat for a small team, searchable internal documents, meeting transcription, a voice interface, read-only tools for a coding client, or a bounded agent task. Start with the smallest stack that supports the outcome, then measure usefulness, reliability, and resource use.

Write down the success criteria before adding services. Identify the users, permitted data, local or hosted model route, acceptable latency, expected concurrency, recovery target, and person responsible for maintenance. A simple deployment with known limits is more valuable than a broad platform whose failures cannot be explained.

Extend the stack safely

A new capability should enter through a clear service boundary with explicit data flow, credentials, network access, persistent state, health checks, and recovery. Prefer a deterministic workflow until model-selected steps provide a demonstrated benefit. Apply authorization before retrieval or action, not as a request to the model after the fact.

Add one dependency at a time and preserve a working rollback point. Test the new component directly before testing the combined workflow. If the change introduces a hosted provider, remote tool, browser credential, writable mount, or new public port, update the threat model and deployment profile at the same time.

Keep commands and versions current

This book explains durable architecture and operating decisions, while the project repositories carry current commands, images, configuration, and compatibility notes. Before acting on a version-sensitive example, compare it with the maintained project documentation.

For an upgrade, record the source commit, effective Compose files, image references, selected models, local overrides, and last successful recovery test. That record turns an evolving open source stack into a deployment you can reproduce and diagnose.

Use the companion repositories

The Self-Hosted AI Stack repository (github.com/hwdsl2/self-hosted-ai-stack) provides integrated Compose configurations. The deployment steps link to focused repositories and API references for InferCrate, GatewayCrate, EmbedCrate, ParseCrate, ScribeCrate, ScribeCrate Live, SpeakCrate, and UplinkCrate. Use the book to understand how they fit together and the repositories to follow their current implementation.

When behavior differs from the book, begin with the relevant repository's current documentation, release notes, and issue tracker. Report a book-only correction using the author contact information in "About the Author."

Build without surrendering control

Control is not a claim printed on an architecture diagram. It comes from knowing where data flows, which identity can perform each action, what state must survive, and how the system fails. Build one bounded path, verify it, protect it, and keep a way back. That discipline matters more than the number of AI services deployed.

The goal is not to avoid every external service or automate every task. It is to make each boundary deliberate. When you can explain where a request goes, which credentials it uses, what state it changes, and how to recover, you have built an AI platform you control.

Appendix A: Services, Ports, and Data Locations

The values below summarize the maintained Self-Hosted AI Stack Compose configurations (github.com/hwdsl2/self-hosted-ai-stack). Before deployment, render the effective configuration for the selected overlays and verify its ports, volumes, and networks. Host access describes the integrated configurations, not every standalone component Compose file. A standalone file can publish a component on all interfaces unless you narrow its mapping.

A.1 Service reference

Service	Role	Default container port	Typical host access
InferCrate	Local model runtime	11434	Internal in integrated stacks
GatewayCrate	Model gateway and admin UI	4000	Published, or loopback with proxy
AnythingLLM	Browser chat UI	3001	Published, or loopback with proxy
EmbedCrate	Embedding API	8000	Loopback
ParseCrate	Document conversion API	5001	Loopback
ScribeCrate	File speech-to-text API	9000	Loopback
ScribeCrate Live	WebSocket and REST speech-to-text	9090 and 8000	Optional loopback; integrated stacks map REST to host port 8001
SpeakCrate	Text-to-speech API	8880	Loopback
UplinkCrate	MCP tool endpoint and dashboard	3000	Internal by default
PostgreSQL	GatewayCrate and optional vector data	5432	Internal

Service	Role	Default container port	Typical host access
Caddy	HTTPS reverse proxy	80 and 443	Public when proxy mode is enabled
goose, optional Chapter 18 service	Interactive AI agent	None	Internal client with no published port

The goose service is defined in a separate Compose overlay and excluded from ordinary stack startup. Its temporary containers use the internal model and MCP endpoints. The example persists configuration and session data in `ai-tools-goose-home`.

A.2 Authentication reference

InferCrate, GatewayCrate, UplinkCrate, ScribeCrate, ScribeCrate Live, SpeakCrate, EmbedCrate, and ParseCrate use generated persistent credentials in the maintained deployments. AnythingLLM generates an initial administrator password and stores current authentication state in its volume. ParseCrate uses `X-Api-Key`; the other component APIs generally use Bearer tokens. Health endpoints may remain unauthenticated.

ScribeCrate Live REST transcription uses a Bearer token. Non-browser WebSocket clients can use the same `Authorization` header or the `token` query parameter during the handshake. Browser clients can use the query parameter because they cannot add arbitrary headers. Use WSS and prevent token-bearing URLs from being recorded in proxy access logs.

Administrative interfaces and ordinary API routes can use different controls. AnythingLLM authentication protects its browser application, GatewayCrate's master key protects administrative access, and MCPHub configuration controls enabled tools. Restrict each interface according to its authority rather than relying only on the component API key.

A.2.1 OpenAI SDK configuration

Export `OPENAI_BASE_URL` and `OPENAI_API_KEY` for the service being called. These loopback URLs are for host clients. Containers on the same Compose network use the service name and container port. Retrieve the component key with `docker exec <container> <helper> --getkey`.

Service	Host SDK base URL	Key helper
GatewayCrate	http://127.0.0.1:4000/v1	litellm_manage
ScribeCrate	http://127.0.0.1:9000/v1	whisper_manage
SpeakCrate	http://127.0.0.1:8880/v1	kokoro_manage
EmbedCrate	http://127.0.0.1:8000	embed_manage
ScribeCrate Live	http://127.0.0.1:8000/v1	whisper_live_manage

ScribeCrate Live's integrated REST host port is `8001`; standalone uses `8000`. EmbedCrate also accepts a `/v1` base URL. For GatewayCrate applications, use a scoped virtual key when available. If API-key authentication is deliberately disabled, omit curl authentication headers; the OpenAI SDK still requires a nonempty key, such as `OPENAI_API_KEY=unused`.

A.3 Persistent data reference

Volume	Main contents
ollama-data	Models, API key, runtime metadata
litellm-data	Master key and gateway configuration
litellm-db	PostgreSQL data
ai-stack-shared	Generated PostgreSQL password
anythingllm-data	Workspaces, chat, documents, settings, authentication
embeddings-data	Model cache and API key
docling-data	Runtime cache and API key
whisper-data	Model cache and API key
whisper-live-data	Model cache and API key
kokoro-data	Model cache and API key

Volume	Main contents
mcp-data	Gateway key, dashboard state, tool configuration
caddy-data	Certificates, private keys, and ACME state
caddy-config	Caddy configuration state
ai-tools-goose-home	Optional goose settings, extension credentials, sessions, and logs

Key-sharing volumes used between containers are regenerated from authoritative service volumes and normally do not need backup.

A.4 External connection reference

Hosted model providers receive prompts and supplied context routed to them. Remote MCP servers can send tool inputs to services such as code hosting services, search providers, or databases. ParseCrate URL conversion fetches remote resources. Model and image downloads contact their registries or model sources unless an offline mode and populated cache are used.

A.5 Stack composition reference

Stack	Main application services
Complete	InferCrate, GatewayCrate, PostgreSQL, AnythingLLM, EmbedCrate, ScribeCrate, UplinkCrate
Chat-only	InferCrate, GatewayCrate, PostgreSQL
Chat UI	Chat-only plus AnythingLLM
RAG pipeline	InferCrate, GatewayCrate, PostgreSQL, EmbedCrate
Full RAG pipeline	RAG pipeline plus ParseCrate
Voice pipeline	InferCrate, GatewayCrate, PostgreSQL, ScribeCrate, SpeakCrate
Voice chat	Voice pipeline plus AnythingLLM
Code assistant	InferCrate, GatewayCrate, PostgreSQL, UplinkCrate, EmbedCrate
AI tools	InferCrate, GatewayCrate, PostgreSQL, UplinkCrate

Chapter 18 can add goose to AI Tools through an optional profile. The agent is not part of the default AI Tools service set.

Optional or commented services can change the effective composition. Render the selected configuration rather than treating this table as executable truth.

For every effective configuration, record whether each published mapping uses all interfaces, loopback, or a specific address. Then test from both the host and another machine. A rendered mapping is configuration evidence; an external connection test confirms the boundary that clients actually see.

Appendix B: Deployment and Recovery Checklists

Copy these checklists into an operations record for each deployment. Add the date, operator, source commit, Compose files, and result. A checked box without evidence is not a substitute for a rendered configuration, application-level test, or restore exercise.

B.1 Deployment profile worksheet

Field	Planned value	Measured value or evidence
Workload and users		
Host location and administrator		
Trust and data boundary		
CPU architecture and core count		
RAM and available headroom		
GPU, VRAM, driver, and runtime		
Storage available and expected growth		
CPU or CUDA Compose path		

Field	Planned value	Measured value or evidence
Enabled and disabled services		
Starter models		
Model artifact size and loaded-weight memory		
Context length and KV cache memory		
Highest tested concurrency		
Published ports and client access path		
External providers and remote tools		
Backup destination and recovery owner		
Maintenance and update window		

After deployment, add idle and peak memory, model storage, representative latency, highest tested concurrency, and the date of the most recent recovery test. Do not copy performance figures from another host into the measured column.

B.2 Pre-deployment checklist

- Define the use case, users, data boundary, and local or hosted model policy.
- Confirm architecture, memory, storage, network, and optional GPU requirements.
- Choose the smallest suitable stack and record its source version.

- Render Compose and inspect images, ports, volumes, networks, and dependencies.
- Plan persistent storage, backups, credentials, and update ownership.
- Start privately, wait for readiness, run the stack check, and test one real path.

B.3 Model evaluation worksheet

Use representative inputs that you are permitted to process. Evaluate the smallest plausible model before allocating more hardware to a larger one.

Field	Candidate A	Candidate B
Exact model identifier and source		
Artifact format, quantization, and file size		
Context setting		
Test task and input set		
Required output criteria		
Quality result and failure examples		
First-response and total latency		
Peak RAM or VRAM		
Highest tested concurrency		
License and permitted use checked		

Field	Candidate A	Candidate B
Decision and reason		

B.4 Internet exposure checklist

- Identify the minimum public entry point and bind other ports to loopback.
- Verify service and administrator authentication, including older installs.
- Configure DNS, HTTPS, request limits, timeouts, streaming, and WebSockets.
- Restrict admin interfaces, tools, mounts, tokens, databases, and outbound access.
- Test the public surface from another machine and review logs for leaked data.
- Document hosted providers and every path that can send data off the host.

B.5 Pre-upgrade checklist

- Read the stack changelog (CHANGELOG.md) and the release notes in each affected component repository.
- Record source commit, image references, Compose files, and local overrides.
- Export keys securely and create a recoverable backup.
- Verify archive checksums and the latest restore exercise.
- Render the new configuration before recreating services.
- Define the rollback images and matching pre-upgrade data set.

B.6 Backup verification checklist

- Archives cover every irreplaceable and key-bearing volume.
- PostgreSQL was stopped or captured with a consistent database dump.
- Backup files have checksums, timestamps, version metadata, and encryption.
- At least one copy is stored on a separate system, and its encryption key is protected separately.

- A clean restore preserves credentials, data, models, and application behavior.
- The restored system can create a new backup.

B.7 Retrieval evaluation worksheet

Record results by question, not only as one average score. A useful set includes answerable questions, ambiguous questions, questions whose answer is absent, and content the current user is not authorized to retrieve.

Field	Result
Corpus version and evaluation date	
Embedding model and vector dimension	
Chunk size and overlap	
Retrieval count and reranking mode	
Expected source for each question	
Expected source retrieved	
Unsupported answer refused or qualified	
Citations point to retrieved evidence	
Unauthorized content excluded	
Latency and failure notes	

B.8 Agent permissions and stop-condition checklist

- State why model-selected iteration is needed instead of a fixed workflow.
- Run the agent under a dedicated identity with only required filesystem and network access.

- Use a restricted model credential and separate credentials for each tool boundary.
- Disable extensions and tools that the current task does not require.
- Classify enabled tools as always allowed, approval required, or denied.
- Require immediate approval for consequential or irreversible actions.
- Set turn, repeated-call, elapsed-time, retry, and cost limits.
- Stop on ambiguous targets, authentication failures, repeated errors, or a request for authority outside the original task.
- Test prompt injection, deceptive tool output, partial failure, retries, and attempts to reach unapproved resources.
- Record sanitized actions and final state, and test how to disable the agent's credentials and tool access.

B.9 Recovery and migration checklist

- Preserve current target data before overwriting any volume.
- Match host architecture, runtime, source, images, and selected configuration.
- Restore authoritative volumes and database state before dependent services.
- Keep public traffic disabled until local functional tests pass.
- Verify keys, database records, models, chat data, tools, optional APIs, and TLS.
- Record the recovery result, remaining exceptions, and safe cleanup date.

Glossary

Agent runtime: The application that operates an AI agent's planning loop, state, tools, limits, and approval policy.

AI agent: A system in which a model can select and repeat actions while pursuing an objective. Its authority still comes from the credentials, tools, network, and files available to its runtime.

API key: A secret value presented by a client to authenticate an API request. Possession of a key does not necessarily identify an individual user or authorize every requested action.

Application programming interface (API): A defined interface through which software clients make requests to a service.

Authentication: Verifying a claimed identity or possession of a credential. Authentication establishes who or what is making a request, but does not by itself determine what that requester may do.

Authorization: Deciding which resources or actions an authenticated user, client, or service may access. Authorization must be enforced at the relevant data and action boundaries.

Bearer token: A credential that grants access to whoever presents it in a request. Protect bearer tokens in storage and transit, and avoid placing them in source code, logs, URLs, or other unintended records. If a client requires a token-bearing URL, use a narrowly scoped credential and redact the URL.

Chunk: A bounded section of source content stored and retrieved as a unit.

Cold backup and hot backup: A cold backup captures data while the affected services are stopped, which simplifies consistency. A hot backup runs while services remain active and requires an application-consistent method, such as a database-native export.

Container: An isolated runtime created from a container image and supplied with one or more processes, runtime configuration, networks, and storage mounts. Replacing a container should not remove data kept in external persistent storage.

Container image: A packaged filesystem and metadata used to create a container. Persistent application data should normally live outside the replaceable container layer.

Context window: The amount of input and generated content a model can process in one request. A supported maximum does not guarantee useful quality or acceptable resource use at that size.

Credential: Information or cryptographic material used to establish identity or access, such as a password, API key, token, or certificate. Protect the secret-bearing parts according to the access they grant.

CUDA: NVIDIA's GPU computing platform, used here for supported accelerated inference images.

Docker Compose: A tool and configuration format for defining and operating related containers, networks, volumes, and service dependencies.

Embedding: A numeric vector representing features of an input such as text.

Endpoint: A network address and path at which an API or service accepts a particular class of requests. A working health endpoint does not prove that other endpoints are authenticated, ready, or functioning end to end.

Gateway: A service that provides a stable entry point and routes requests to one or more backends.

GGUF: A file format used by llama.cpp-family runtimes to store model tensors and loading metadata. GGUF does not itself identify a model's architecture, parameter count, quantization, or quality level.

Graphics processing unit (GPU): A processor designed for highly parallel computation. Supported AI runtimes can use a compatible GPU to accelerate model inference and other workloads.

Health check: A test that reports whether a process or service satisfies a defined condition. Liveness, readiness, and end-to-end functionality are different conditions.

Hosted provider: An externally operated service that supplies model inference or another AI capability. Data sent to it crosses the local trust boundary and becomes subject to the provider's availability, pricing, terms, and data practices.

Inference: Running a trained model to produce an output from an input.

Ingestion: Preparing source data for retrieval, commonly by parsing, normalizing, chunking, embedding, and storing the resulting text and metadata.

JavaScript Object Notation (JSON): A text format for structured data using objects, arrays, strings, numbers, Boolean values, and null. Many APIs in this book accept or return JSON request and response bodies.

Key-value cache (KV cache): Runtime memory that retains attention keys and values for tokens in an active model context so they do not have to be recomputed for each new token. Longer contexts and parallel sequences can increase its memory use.

Large language model (LLM): A model trained to process and generate language.

Least privilege: Granting only the access required for a specific task.

Loopback: A host-local network interface, commonly represented by `127.0.0.1`. Binding a published port to loopback prevents direct access from other machines, although local processes can still reach it.

Metadata: Descriptive or structural information associated with a record, such as its source, page, owner, timestamp, or content type. Retrieval systems can use metadata for filtering and attribution. Trusted metadata can also support server-side authorization filters.

Model alias: A stable client-facing model name that a gateway maps to a specific local or hosted provider route.

Model Context Protocol (MCP): A protocol through which compatible clients can discover and call tools or access contextual resources.

Model token: A unit of text produced by a model's tokenizer. A token is not necessarily a whole word, and input and output token counts affect context limits, latency, and sometimes cost. It is distinct from an authentication token.

Named volume: Docker-managed persistent storage that exists independently of an individual container. Removing a container normally preserves its named volumes, but explicitly deleting those volumes removes their stored data.

OpenAI-compatible API: An API that implements selected OpenAI request and response conventions so compatible clients can use another model backend. Compatibility is specific to the implemented endpoints, fields, streaming behavior, and API version rather than a guarantee of identical features.

pgvector: A PostgreSQL extension that adds vector data types, distance operators, and indexes for similarity search.

Prompt injection: Untrusted content that attempts to alter a model-driven system's instructions or induce unauthorized behavior. It must be addressed by application policy and permissions, not only by another prompt.

Quantization: Representing model weights or computations with reduced precision. For model weights, this generally reduces storage and loaded-weight memory, with quality, performance, and compatibility tradeoffs that depend on the model, method, runtime, hardware, and workload.

Representational State Transfer (REST) API: An API designed according to the REST architectural style, commonly using resources, HTTP methods, and status codes. In this book, REST API refers to ordinary HTTP request-response endpoints rather than persistent WebSocket sessions, without claiming that every service implements every REST constraint.

Reranking: Rescoring a retrieved candidate set against the original query to improve its order.

Retrieval-augmented generation (RAG): Supplying a generative model with context retrieved from an external collection.

Reverse proxy: A service that accepts client traffic and forwards it to an internal application, often adding TLS and traffic controls.

Self-hosted: Operated on infrastructure controlled by the deployer. It does not by itself guarantee privacy or isolation.

Semantic search: Retrieval based on the meaning of a query and candidate content, commonly implemented with embeddings and vector similarity. Similarity does not by itself guarantee relevance, factual accuracy, or authorization.

Speaker diarization: Determining which speech segments came from different speakers. Diarization labels speakers within a recording but does not identify their real-world identities unless a separate recognition system is used.

Speech-to-text: Converting spoken audio into text through batch or streaming transcription.

Text-to-speech: Generating spoken audio from written text.

Threat model: A structured description of assets, actors, trust boundaries, threats, assumptions, and controls used to reason about a system's security.

Tool: A bounded capability that a client or runtime can invoke, such as reading a file or querying a database.

Transport Layer Security (TLS): The protocol used by HTTPS and secure WebSockets to encrypt traffic and authenticate the server and, optionally, the client.

Vector: A fixed-length numeric representation used here to compare the semantic similarity of text.

Vector database: A data store that supports similarity search over vectors and associates them with records and metadata.

Video memory (VRAM): Memory attached to a GPU and used for model weights, runtime caches, and accelerated computation.

Virtual key: An application credential created by LiteLLM with optional model, budget, and expiration restrictions, distinct from the administrative master key.

Voice activity detection: Detecting speech and non-speech regions in audio. It can help segment streaming input and avoid processing extended silence, but it does not transcribe speech or identify a speaker.

WebSocket: A persistent, bidirectional connection used here for streaming audio and real-time transcription events. Secure WebSockets use the `wss://` scheme.

About the Author

Lin Song, PhD, is a software engineer, author, and open source developer. He creates practical tools for self-hosted AI, online privacy, and secure infrastructure, including the Self-Hosted AI Stack and Setup IPsec VPN projects. Since 2014, his VPN projects have collectively earned 40,000 GitHub stars, while their Docker images have been pulled more than 30 million times, helping millions of users build and manage their own VPN servers.

His self-hosted AI projects provide containerized deployments for local models, private chat, document intelligence, voice pipelines, and MCP tools. His other books cover online privacy and building IPsec, OpenVPN, and WireGuard servers. His work combines maintained open source projects with clear, practical guidance for building and operating systems you control.

Connect with Lin Song

- Amazon author page (amazon.com/author/linsong)
- GitHub (github.com/hwdsl2)
- LinkedIn (linkedin.com/in/linsongui)

A Note from the Author

Thank you for reading. I hope this book helps you build a self-hosted AI platform that is useful, understandable, and maintainable. If you found the book helpful, an honest rating or short review would be greatly appreciated.

Lin Song
Author

Index